SOCIAL CAPITAL AND
URBAN NETWORKS OF TRUST

Social Capital and Urban Networks of Trust

Edited by

JOUNI HÄKLI
University of Tampere, Finland

and

CLAUDIO MINCA
Royal Holloway, University of London, UK

Routledge
Taylor & Francis Group
LONDON AND NEW YORK

First published 2009 by Ashgate Publishing

Published 2016 by Routledge
2 Park Square, Milton Park, Abingdon, Oxfordshire OX14 4RN
711 Third Avenue, New York, NY 10017, USA

First issued in paperback 2016

Routledge is an imprint of the Taylor & Francis Group, an informa business

British Library Cataloguing in Publication Data
Social capital and urban networks of trust
1. Social capital (Sociology) 2. Trust 3. Economic development 4. Community life - Italy - Case studies
5. Community life - Finland - Case studies
I. Häkli, Jouni II. Minca, Claudio
307.7'6'094

Library of Congress Cataloging-in-Publication Data
Social capital and urban networks of trust / [edited by] by Jouni Häkli and Claudio Minca.
p. cm.
Includes bibliographical references and index.
ISBN 978-0-7546-7319-4
1. Social capital (Sociology) 2. Trust. 3. Economic development. 4. Community life--Italy--Case studies. 5. Community life--Finland--Case studies. I. Häkli, Jouni. II. Minca, Claudio.
HM708.S623 2008
307.76094--dc22
2008039487

ISBN 13: 978-1-138-24545-7 (pbk)
ISBN 13: 978-0-7546-7319-4 (hbk)

Contents

List of Figures and Tables

Figures

Tables

Acknowledgements

This volume originated in part from a collaborative research project, coordinated by Jouni Häkli (as Director) and Claudio Minca (in charge of the Italian part of the project), funded by the Academy of Finland. While six of the chapters are the outcome of this project, it is important to emphasize that this book in no way represents a 'research report', but intends to provide a major statement on its own. We would like to thank the Academy of Finland for financial support (Grant SA 105132), as well as the supportive research environment provided by the Institute for Social Research at the University of Tampere and the Department of Geography at Royal Holloway, University of London. In this respect, we also wish to thank the research program on Social Capital and Networks of Trust (SoCa) of the Academy of Finland 2004-2007, coordinated by Dr Pertti Jokivuori in the University of Jyväskylä, Finland. Special thanks are also due to the Department of Regional Studies at the University of Tampere for financially supporting the language editing of this book. Finally, we wish to extend our gratitude to all our contributing authors and especially to Lowri Jones and Ben Coles for their precious editorial assistance. We owe the completion of this volume to all of you.

Jouni Häkli and Claudio Minca
Tampere and London, June 2008

Introduction

Jouni Häkli and Claudio Minca

Research on social capital and networks of trust expanded rapidly after the late 1980s and now comprises a broad and heterogeneous field of scholarship that spans most social science disciplines, from anthropology to political science, from sociology to human geography (see Coleman 1988; Fukuyama 1995; Nahapiet and Goshal 1998; Portes 1998; Mohan and Mohan 2002; Holt 2008). One of the most influential exponents of the notions of social capital and trust has been Robert Putnam, whose research on civic traditions in Italy and the United States rapidly became the pacesetter in the field (Putnam 1993; 2000). Importantly, Putnam's influence has not been limited to the promotion of social capital as an analytical concept; social capital has also become a key policy tool. This policy interest has not been confined to Italy, the US and other Western industrialized countries since, with the support of the World Bank and other international organizations, the idea of social capital as a remedy to various social ills has also been adopted in a large number of developing countries (see Woolcock and Narayan 2000; World Bank 1997).

Arguably, the notion of social capital has become fashionable both in academic and policy-making circles, resulting in a rapid proliferation of publications engaging with the concept both theoretically and empirically – sometimes with notably normative overtones (Putnam 2000). In this extensive literature the notions of social capital and trust have been approached from very different points of view and treated with various levels of generalization, so that at times the two notions have come notoriously close to being 'chaotic concepts' (Martin and Sunley 2003). This tendency has not only produced incoherent and muddled conceptualizations but has also often rendered social capital research subject to purposeful 'misuse' in legitimating neoliberal politics and policy-making – a tendency discussed further by Mauro Cannone in the second chapter of this volume (see also Levi 1996; Tarrow 1996; Fine 2002).

Critically aware of these problems and related risks, this collection of essays seeks to assess in depth how trust and social capital influence the ways in which networks of actors are constituted and function in different urban localities. Unlike the approach popularized by Putnam and his followers, our approach is deliberately critical of assumptions concerning the role of social capital as an unproblematic collective resource, 'magically' available to all. Instead, we adopt an understanding of social capital as an unevenly distributed resource that depends on individuals' ability to enact the power potentials that reside in their membership in social networks. This conception, originally proposed by Pierre Bourdieu (1980; 1986), sets the tone of this volume. Moreover, we contend that besides formal decision-making there are significant informal arenas through which trust is negotiated and immaterial capital is created, and that it is these arenas and practices that often shape local development processes (see also Banks et al. 2000; Lee et al. 2005).

Our purpose, then, is to advance a theoretical understanding of the historical and cultural embeddedness of these processes in several urban contexts in Italy and in Finland. The focus on these particular national and urban contexts has several theoretical and empirical foundations. Firstly, Italy is the place where Putnam famously 'discovered' social capital and where the functioning of social capital is normally assumed to be crucial to any political and economic endeavour. We thus thought that challenging the ways in which social capital is presented here, the site of its 'original' theorization, both at the institutional level and in the media, was a particularly intriguing project. Secondly, Italy and Finland represent, in many ways, two paradigmatic cases where the role and the functioning of trust and social capital are expressed in extremely different ways: often informal and almost cryptic in Italy; highly formalized and institutionalized, even planned and sometimes promoted, in Finland. Examining these two very differentiated contexts, we thought, could prove useful in the understanding of *social capital in place* and for reflecting on its theoretical and practical implications. The Italian and Finnish cases discussed in this book should thus be seen as tentative laboratories for the exploration of the place-based nature of the enactment of social capital. The third reason has to do with the fact that, in the light of the previous considerations, we decided to actually write a comparative research project on the functioning and the nature of social capital in Italy and Finland. Some of these essays and, possibly, the very idea of this book stem from that ongoing project and should be read as a contribution to a theoretical 'work-in-progress' on the spatializations of social capital.

This volume seeks therefore to contribute to the body of knowledge concerning the role of urban cultures (context-specific meanings, identities, norms, values, structures of expectations, and patterns of behaviour) in local economic and political performance. More specifically, the book hopes to contribute to the understanding of why disparities in urban socio-economic development persist and are, indeed, increasing in the contemporary globalizing condition, characterized as it is by seemingly boundless flows of people, economic capital and information (Brenner and Theodore 2002; Martin and Sunley 2003). Contributions in this volume chart the intersection of three areas of interest: the (re)production and functioning of trust, the existence and uses of social capital, and the transformations that urban localities are undergoing in the context of political and economic globalization. To this end, the case studies presented here examine the role of social capital in the constitution and reproduction of urban networks of trust and, more specifically, how social capital and trust are reflected in the capacity of these networks to achieve their goals and to deliver specific forms of urban development. The volume also puts forth theoretical work tying the key concepts of trust, urban networks and social capital explicitly into society's institutional structures (see also Maloney et al. 2000; Adam and Roncevic 2003). Each contribution argues for the importance of place and emphasizes how the functioning of urban networks necessarily reflects the particular historical, cultural and political contexts of the locality in question.

While not organized in explicitly comparative fashion, the individual case studies in this volume do, nonetheless, enable cross-cultural comparisons between cities with significant differences in terms of regional identity, cultural traditions and social capital embedded in urban networks. The chapters also reflect differences in how trust is established between actors, and in the ways in which new social actors can become members of established local networks. They also show that urban networks may be more or less place-based; that is, the principles of their formation and functioning may encourage more or less socially and spatially open membership based on the actors' sense of the locale and on their capability of interacting in a variety of socio-spatial contexts.

By addressing the questions of network membership, trust and policy performance this edited collection attempts to trace the invisible resources that influence the productivity of networks and organizations. In geographical terms, the contributions shed light on some important ways in which 'place' and 'locale' affect the negotiation of socio-economic development processes in different kinds of urban contexts. To this end the case studies trace those cultural, political and institutional aspects that influence trust in actor networks, as reflected in the economic and political successes or failures of specific urban development projects. Particular emphasis is placed on the constitution and enactment of social capital within each project.

The volume is composed of ten chapters organized into three parts. The first part, 'Re-placing Social Capital', contains three contributions that focus on the conceptual and political ramifications of trust, social capital and local economic development from a critical geographical point of view. These introductory chapters are followed by three original case studies that examine how social actors establish *grounding networks* for achieving political and economic goals that would not be attainable otherwise. The third part of the volume (made up of four chapters) explores the *mobilization of trust* in such networks, so as to illuminate the constitution and enactment of social capital in place.

Re-placing Trust and Social Capital

In the opening chapter Jouni Häkli examines the multiple ways in which geographical aspects of trust have figured, but nevertheless remained unaccounted for, in the mainstream social science literature on the topic. In what, necessarily, remains only a tentative assessment of this extensive literature, the chapter reflects on three distinct geographies of trust, each of which may facilitate a critical re-examination of the significance of the concept for human geography and other social sciences. The first is what Häkli calls a *paradoxical* geography of trust, referring to the ubiquity of the term as both an analytical tool of political philosophy and a discursive constituent of Western modernity. He argues that all attempts at empirically 'locating' or 'finding' trust in social practices related to economic transactions, democratic government and policy formation are bound to 'succeed' because trust is considered as something 'always already there' as a generally accepted means of making sense of society and thus a generic part of Western modernity. Secondly, Häkli points out a *hidden* geography of trust by exposing spatial assumptions concerning the form and functioning of various communities in which trust is empirically or theoretically analysed. Thirdly, the chapter discusses the ways in which a *reified* geography of trust has emerged through more than five decades of empirically based research on Italian civic culture. In all, the chapter makes visible some of the spatial assumptions that form the background of most analyses of thought on trust, arguing that these are not merely floating geographical imaginations, but fundamental systems of signification that influence our perception of social processes, the shape and nature of political communities, and the constitution of trust relationships.

Extending the discussion explicitly towards the concept of social capital, Mauro Cannone addresses in the second chapter the trajectory taken by the ideas by Robert Putnam and his followers, from their original conceptualization to 'the corridors of power' in contemporary Western societies. Used as an independent variable to explain not only political performance but also the economic success of a number of localities at various scales of analysis, social capital was envisaged in Putnam's research as a by-product of these localities' predisposition towards cooperation and association, stemming from specific civic attitudes rooted in some places but not in others. However, critical of ways in which the concept has been used not only in academic research but also in political and policy-making contexts, Cannone argues that ideas of social capital have, in fact, served to legitimate contemporary neoliberal policies related to the dismantling of the welfare state by downsizing public expenditure. To this end Cannone places the 'Putnamian' understanding of social capital within a broad and complex historical and political framework and shows how, in providing a mono-causal explanation for a broad range of socio-economic phenomena, such as the political and economic 'success' of countries, regions and cities, this powerful discourse has helped to conceal rather than expose the structural roots of the fundamental socio-economic changes of the last decades. Cannone concludes that in applying a neoliberal logic to their idea of social capital, Putnam and his followers end up uncritically embracing these new conditions without ever discussing the relations between negative or positive trends in social capital and their related political implications.

In his chapter, Paolo Giaccaria explores the role of social capital in the vocabulary of local and regional development. By selectively scrutinizing the Italian and the international literature on local development and regional competitiveness as a case in point, he shows that although the concept began as an enthusiastic recognition of the social dimension of economic life, it has paradoxically become a powerful tool in the 'economization' of social and political life. Drawing a parallel between discourses on social capital and those seeking to understand the role of the 'locale' in economic development, Giaccaria argues that both approaches tend to suffer from methodological ambiguity and vagueness. As a consequence, they tend to grossly oversimplify the complexity – and the place-based nature – of the social processes that lead to 'economic development', translating into a kind of 'verbal magic', which should instead be rigorously analysed. Moreover, Giaccaria suggests that this 'magic' – the lack of a rigorous methodological foundation – can easily become instrumental in promoting elite views as the necessary and natural consensus expressed in holistic discourses about social capital and territorial development. Such marginalization of dissident voices results in the loss of 'the political' and the consequent replacement of the local social and cultural dialectic with the 'Putnamian' ideal of a trustful local community where individual interests blur into shared stake-holding and collective agency. As a case in point, Giaccaria refers to discourses of urban and regional competition that have forcefully highlighted demands for consensus and cohesion, implying that the whole territory should be organized in a particular way for economic purposes so as to be competitive, and downplaying the conflictual dimensions of the interaction among different groups within localities. In what the author calls a 'magic and loss *dispositif*', conflict is then pushed outside the boundaries of the local community and, instead, located in competition against other territories, while the 'local' is praised as the locus of consensus and cohesion.

Grounding Networks

Moving on to discuss more concretely the ways in which social capital is enacted in urban networks, Giuseppe Porcaro and Claudio Minca address the case of Trieste's failed bid for the Expo 2008. In the light of this failed project, the authors set out to explore the spatial dimensions of social capital and trust in terms of the capacity of the local networks related to the Expo 2008 bid to mobilize trust relationships and a long-standing tradition of local 'working togetherness' – elements that are often identified as the outcome of the enactment of strong, well established collective social capital. To this end, Porcaro and Minca examine the ways in which a series of historical and spatial metaphors have been used to put Trieste 'on the map' and how these very metaphors are associated with the city's prestigious past in the bidding documents, the official presentations of the city's candidature, and the geographical imaginations mobilized by the local media. The authors argue that the failure of the bid was the consequence of an underlying conflict between the image of Trieste as promoted by the bid and the actual political and economic situation of the Adriatic city today. An equally problematic shortcoming that the authors point out was the modest and passive involvement of the local population in the initiative – a critical element of support in any competition of this sort, as highlighted also in the chapter that follows.

Poutiainen and Häkli examine the role of place and vision in building trust and engagement among participants in a major urban development project in Tampere. They begin by outlining how a particular narrative of Tampere as a city on the leading edge of technological progress emerged from the late nineteenth century onwards, and was then revitalized as part of the official vocabulary revolving around the eTampere programme. The authors show how this 'progressive' identity of Tampere was successfully inscribed into particular visions and goals that materialized in the ambitious development programme and helped shape its contents and contours. By foregrounding the idea of urban narratives as a 'glue' in social interaction, Poutiainen and Häkli explore the way in which these stories function as a form of cognitive social capital, which builds a sense of togetherness among heterogeneous actors and directs the processes of recruiting and engaging new members into the programme network. By means of actor-network theory, they discuss the concrete formation of the (eTampere) network in terms of a process of 'translation', where social capital became a critically important but scarce resource when the (eTampere) network strove to expand beyond the core members directly involved in the programme's design. Consequently, a broader participation was only partially reached, and the fringes of the programme network became unstable and porous, suffering from a notable lack of public support. Where the programme turned out to be more successful was in its ability to mobilize a vision of Tampere as an intellectually progressive industrial centre, and with eTampere as its 'natural' extension. While failing in its grandiose goal of boosting the city into a world-leading position, the programme nevertheless managed to amass under its conceptual umbrella hundreds of information and communications technology projects.

The role played by narratives and discourses on local history also takes centre stage in the chapter by Stefano Soriani, who analyses the setting up and the consolidation of cooperative social networks related to attempts to address the 'Venice problem' and the 'place' of the port of Venice in the future of the city. Ever since the dramatic flooding of the 1960s and the implementation of the UNESCO initiatives of the 1970s for 'preserving and saving' the city, Venice has witnessed a wide-ranging debate about its future. Grounding his approach in these broader considerations, Soriani focuses in particular on the port of Venice, which has been the driving economic force in the city for many decades. The author explores some of the ways in which the port sector (both the Port Authority and a variety of private firms which operate in the port) contributes to the realization of networks of actors with significant influence on the city's urban policies – as well as its visions for the future. By scrutinizing technical reports, previously executed studies, research projects, comments in the local and national press, and 'texts' that give voice to the principal actors involved, Soriani shows how the port itself has helped shape both the context of the 'Venice problem', and the formation of networks that seek to identify, re-interpret, and put into practice new resources for the revitalization of the port of Venice, in the broader context of urban economic and political transformation of the city. With particular emphasis on the formation and the consolidation of a new Port Business Community, Soriani focuses on the part played by narratives about the port's capacity to bring about development by sustaining the mobilization of important resources. In this regard he suggests that trust and distrust are dynamic factors that can contribute to re-contextualizing the 'local' by promoting the formation of new networks, social legitimacy and, eventually, social capital.

Mobilizing Trust

In assessing how and under what conditions trust may be established between actors in urban networks, Claudio Minca sets out to trace the historical, cultural and political coordinates that have made the use of the *triestino* – a specific dialect spoken in Trieste – the privileged space for the enactment of Triestine social capital. On the basis of a meticulous and thorough account of the historical evolution of *triestino*, Minca argues that in order to understand the functioning of social capital in Trieste it is necessary to take into account the crucial role played by the dialect in the material and, especially, immaterial exchanges based upon trust. As a spoken lingua franca this dialect is based on shared practices nurtured in personal relationships and social and economic networks that activate a shared set of understandings, ways of saying and ways of doing 'what needs to be done'. Precisely because *triestino* is the language of practice, it continues to play a key role in defining the relationships that matter, thus acting as the prime vehicle in mobilizing trust. Minca argues then that the enactment of social capital in Trieste usually passes through a series of informal practices that materialize within both real and imagined places. Triestine social capital thus not only expresses itself in a series of spatial practices but activates itself through a specific language which, in that particular moment and place, makes possible a 'connection' between the individuals, the city and its memory, determining the success (or failure) of future projects.

With emphasis on the role of the visual and the textual rather than oral or spoken communication, Häkli explores a case of trust building in the context of an ambitious development project called *På Gärnsen-Rajalla* carried out by the twin-town of Haparanda-Tornio. Located at the borderland between Sweden and Finland, this extraordinarily challenging project aims at building a transnational space by (physically) uniting the centres of two towns located in two different countries (Sweden and Finland). Häkli approaches the problem of transnational trust with conceptual tools from actor-network theory, focusing on the manner in which trust is related to the boundary, which acts both as a force of divergence and a point of contact in international networks of cross-border cooperation. Foregrounding the Tornio River as a set of 'boundary objects', the chapter explores cross-border cooperation between Haparanda and Tornio, and particularly the *På Gränsen-Rajalla* project, seeking traces of the river as an element that helps build trust amongst the participants of cooperative networks as well as within the broader population. Häkli concludes that, while the work in the *På Gränsen – Rajalla* project is still ongoing, it is already clear that the Tornio River has fostered trust-building amongst the key actors of the cooperative network who have then been able to successfully mobilize resources for overcoming the social, cultural and material complexities that pertain to the production of transnational space.

In his chapter on a specific case of successful enactment of social capital in Venice, Mauro Cannone examines the dynamics of urban regeneration in a context where development strategies have had to face exceptional challenges set by the built environment. Focusing on the redevelopment of the Arsenale, a vast semi-abandoned area formerly hosting industrial and military activities and now lying as a problematic urban void at the heart of Venice, the chapter explores the process of the constitution of the Maritime Technologies Centre (*Thetis*) located in this area. The author argues that, besides contributing to the rejuvenation of the Arsenale, the project has also helped to revitalize the socio-economic life of a city struggling against the permanent loss of residents and services. Rather than framing *Thetis* as the result of a coherent strategy put forward by urban administrators and business leaders, Cannone traces its constitution by looking at how a more precarious network of actors was successfully formed around the project, capable of mobilizing resources embedded in the local milieu of Venice. Recognizing the significance of the presence 'in place' of specific resources, and the ability to mobilize and make use of them, Cannone's chapter suggests that urban development is differently embedded in every local milieu, and that it is precisely by means of a series of highly differentiated 'social' relations that such locally embedded preconditions can be transformed into key resources, or 'capital', for development.

With emphasis on voluntary collective action, the chapter by Martti Siisiäinen examines the role of associational social capital embodied in the field of welfare policy practices in the city of Jyväskylä. Siisiäinen understands the establishment of voluntary associations in terms of organized social capital, a process where networks are formalized, participation is tied to membership roles, and activities occur mostly autonomously from individual members. The author approaches voluntary associations as a form of social and symbolic capital by first contextualizing their role within a typology of European polities. He then analyses the formation and characteristics of registered associations in Jyväskylä with attention to their differentiation since 1919. By discussing the changing role of voluntary associations in the political system of welfare provision, Siisiäinen examines more generally the ongoing transformations within associational social capital in the city of Jyväskylä. As a general conclusion, he points out that the borderline between civil society and local government runs not between but *inside* associations, so that those actors who interact closely with the local government tend to function as its 'accessories', while 'rank and file actors' remain in their civil society roles. Moreover, since an active associational field is a potential source of innovation and key to the development of local democracy, it is important that both conflicting and integrative movements and associations are present and active in any local society.

We believe that the itinerary traced by this volume through different urban laboratories for the production of social capital and trust in Italy and Finland is particularly useful in shedding light on some elements that the conventional debate on social capital tends to overlook. The radically distinct understanding and workings of social capital in these two national contexts reveals, in fact, the importance of matching a more formal, so to speak, 'institutional' reading of social capital with the recognition of the crucial importance of its informal manifestations. This realization presents, of course, a set of methodological problems that are typical of any investigation of 'communities of practice', based as they are on a conceptual as well as very 'practical' dimension: dimensions that tend to merge and become indistinguishable when social capital is actually enacted and translated into spaces of either collaboration or resistance. This double expression of social capital reveals how important it is to reflect on the cultural and political nature of the production of networks of trust – a production that is always, necessarily, embedded in the 'local' and that cannot therefore be generalized. Despite this, however, what we hope to have achieved with this volume is to show how reflecting on the geographical and, at the same time, (almost) immaterial dimensions of social capital is particularly important in order to understand the ways in which networks of trust operate in and through 'real' places where power, in its relational forms, is enacted by and upon people, and translates this very 'immateriality' into concrete projects that can succeed or fail, but that inevitably help shape the social reproduction of the urban fabric.

Bibliography

Adam, F. and Roncevic, B. (2003), 'Social capital: recent debates and research trends', *Social Science Information* 42:2, 155–83.

Banks, M. et al. (2000), 'Risk and trust in the cultural industries', *Geoforum* 31:4, 453–64.

Bourdieu, P. (1980), 'Les capital social: notes provisoires', *Actes de la recherché en sciences socials* 31:1, 2–3.

Bourdieu, P. (1986), 'The forms of capital', in J. Richardson (ed.).

Brenner, N. and Theodore, N. (2002), 'Cities and the geographies of "actually existing neoliberalism"', *Antipode* 34:3, 349–79.
Coleman, J. (1988), 'Social capital in the creation of human capital', *American Journal of Sociology* 94, 95–121.
Fine, B. (20042), 'They f**k you up, those social capitalists', *Antipode* 34:4, 796–9.
Fukuyama, F. (1995), *Trust: The Social Virtues and the Creation of Prosperity* (New York: Free Press).
Holt, L. (2008), 'Embodied social capital and geographic perspectives: performing the habitus', *Progress in Human Geography* 32:2, 227–46.
Lee, J., Árnason, J.A., Nightingale, A. and Shucksmith, M. (2005), 'Networking: social capital and identities in European rural development', *Sociologia Ruralis* 45:4, 269–83.
Levi, M. (1996), 'Social and unsocial capital', *Politics and Society* 24:1, 45–55.
Maloney, W., Smith, G. and Stoker, G. (2000), 'Social capital and urban governance: adding a more contextualized 'top-down' perspective', *Political Studies* 48:4, 802–20.
Martin, R. and Sunley, P. (2003), 'Deconstructing clusters: chaotic concepts or policy panacea?', *Journal of Economic Geography* 3:1, 5–35.
Mohan G. and Mohan J. (2002), 'Placing social capital', *Progress in Human Geography* 26:2, 191–210.
Nahapiet, J. and Goshal, S. (1998), 'Social capital, intellectual capital, and the organizational advantage', *Academy of Management Review* 23:2, 242–66.
Portes, A. (1998), 'Social capital: its origins and applications in modern sociology', *Annual Review of Sociology* 24:1, 1–24.
Putnam R. (1993), *Making Democracy Work: Civic Traditions in Modern Italy.* (Princeton: Princeton University Press).
Putnam, R. (2000), *Bowling Alone: The Collapse and Revival of American Community* (New York: Simon & Schuster).
Richardson, J. (ed.) (1986), *The Handbook of Theory: Research for the Sociology of Education* (New York: Greenwood Press).
Tarrow, S. (1996), 'Social movements in contentious politics: a review article', *The American Political Science Review* 90:4, 874–83.
Woolcock, M. and Narayan, D. (2000), 'Social capital: implications for development Theory, Research, and Policy', *World Bank Research Observer* 15:2, 225–50.
World Bank (1997), *World Development Report 1997* (New York: Oxford University Press).

PART 1
Re-placing Social Capital

Chapter 1
Geographies of Trust

Jouni Häkli

Introduction

Trust is a long-term key word in social theory, if less so in human geography. Its winding conceptual trajectory crosses almost all social scientific disciplines from political science and political economy to sociology to nursing, with partly differing times of prominence in each. Early treatment of trust as a part of responsible government can be found in the *Second Treatise of Government* by John Locke (1690/1980) for whom it was essential that officials do not act 'contrary to trust' (Parry 1976, 131). The significance of trust as a social condition was noticed in the early eighteenth century by the father of modern economics, Adam Smith, for whom it represents a link between the calculative commercial society and the broader civil society, within which the former was embedded (Porta 1998, 110). Since Locke and Smith numerous classic texts and scholarly works have traced the foundation and functioning of social trust (for example, see Tocqueville 1835/2000; Mill 1848/2006).

Because of its philosophical nature, the concept of trust has received the most theoretical attention at times when anti-positivistic scholarship has gained a foothold in human and social sciences. This holds true especially for sociology, where interest in trust coincides with the rise of the Germanic tradition of culturally sensitive sociological analysis in the classic scholarship by Georg Simmel (1858–1918), Werner Sombart (1863–1941) and Max Weber (1864–1920). Also the latest upsurge of scholarship on trust has gained momentum from the cultural and linguistic turns in social research.

With roots wide and deep in Western thought, it may be difficult to trace the different geographical modalities that pertain to the concept of trust. This difficulty is reflected in the scarce amount of theoretical discussion that exists on the matter even within human geographical scholarship on trust (see Hudson 1998; Mohan and Mohan 2002). While some attempts to conceptualize the spatial ramifications of social trust exist, there are very few works that seek to explore the geographical assumptions that inform theories of trust. Convinced that implicit geographies of trust can be teased out with a closer examination of the intellectual and empirical contexts of the theories of trust, I set out here to explore this understudied field.

My aim is threefold. First, I seek to show that the concept of trust is not the innocent analytical tool it is often taken to be, especially in empirically oriented research, but part and parcel of the discursive constitution of the phenomena that these studies strive to describe and explain. Second, I aim to make visible some of the spatial assumptions that work in the background of analytical thought on trust, arguing that these are not merely floating geographical imaginations, but fundamental systems of signification that influence our perception of social processes, the shape and nature of political communities, and the constitution of trust relationships. Third, along with authors such as Livingstone (1995) and Lorimer (2003), I argue that place matters in the production of knowledge, and that particular geographical contexts of research may give rise to patterns of thought that universalize the particular by transferring interpretative schemes from one context to another without due attention paid to the geography of the process (Häkli 2003; Sidaway 2008).

There are at least three relevant geographies of trust that may be helpful in the critical re-examination of the significance of the concept for human geography and other social sciences. First, there is a *paradoxical* geography of trust, by which I mean the fact that through centuries of classic texts in Western political philosophy, economy and sociology the concept of trust has infiltrated the very processes of democratic government and state formation which constitute the political framework of modern societies. Through the loops of such 'double hermeneutic' reproduction, the geography of trust has become almost generic to Western modernity. The term 'double hermeneutics' is used by Anthony Giddens (1976) in reference to the observation that when scientific concepts become generally accepted as means of making sense of the society, they not only reflect but also construct social practices. Likewise trust not only describes but also construes social practices. Second, I argue that a *hidden* geography of trust can be exposed by pointing out the spatial assumptions that concern the form and function of the political or cultural communities in which trust is empirically or theoretically analysed. Third, I want to expose the manner in which a *reified* geography of trust has emerged through more than five decades of empirically based research that has produced an extensive literature assuming that a certain kind of relationship exists between (Southern) Italian civic culture and political trust.

I begin this chapter by tracing the conceptual roots of trust and thereby seek to show the generic nature of the concept in the Western political modernity. I then focus more on the hidden geographical dimensions of theories of trust and finally explore the manner in which geography of trust has been reified mainly in the context of American political science scholarship on Italy. I conclude by discussing contemporary challenges for the analysis of social trust and propose some avenues for approaching the question in future research.

Why Trust Matters: The Geographical Paradox of Trust

Much of the recent interest on trust can be attributed to the burgeoning research on social capital since the early 1990s. The concept has been vital especially in those strands that follow James Coleman's (1988) and Robert Putnam's (1993; 2000) thought where trust is seen as a resource pertinent to collectives. It is, however, possible to trace a Western political philosophical discussion of the role of trust in civil society that extends back much further. In this discussion 'civil society' has emerged as an important mediating field of social relationships and action between the individual or family and the state. It is this linkage that has been assumed to exist between trust, civil society, and national economy, which serves as the first of our three avenues for approximating the geographies of trust.

In contrast with Thomas Hobbes (1651/2006) and Adam Ferguson (1767/1980), for whom civil society was more or less the polity itself, a line of political philosophical thinking from Georg Wilhelm Friedrich Hegel (1821/1991) to John Stuart Mill (1848/2006) to Antonio Gramsci (1929–35/1988) up to the present day (Rawls 1971; Etzioni 1993) has sought an understanding of civil society in terms of general moral bonds formed between social actors (Porta 1998). These bonds are seen to provide the preconditions for a self-regulating community that exists between individuals, which is, nevertheless, distinct from that of the nation-state (Seligman 1992, 6).

Trust figures into this fundamental theory of 'civilized society' as the indispensable glue that holds the civil society together (Parry 1976, 129). Based on mutuality and communality between moral human beings, trust exists as the social and cultural mechanism that balances between unselfish emotions and rational evaluation, and thus enables private and public interests to coexist (Porta 1998, 110). However, as aptly shown by the complex philosophical discourse on the relationship between trust and civil society, this relation is far from being unproblematic. Ambiguity exists, for instance, on whether a high degree of social trust, and particularly the ensuing consensus among the political elites, is always a precondition for, or conducive to, good government. Also the question of trust-based cooperation as a benevolent condition in all circumstances remains a contested topic (Parry 1976, 130; Gambetta 2000, 214).

Nevertheless, most theoretical accounts of social trust converge on the principle that a trustful civil society is likely to be better than a distrustful one. At the interpersonal level, trust among the citizenry may foster cooperation and beneficial competition necessary for a viable and innovative civil society. At the macro level, social trust is seen as an important factor in nurturing social cohesion, strengthening democratic governance and improving the efficiency of public administration (see Almond and Verba 1963; Putnam 1993). Simply put, trust is considered essential for the functioning of democratic societies for three reasons. First, the mutual communication required in processes of democratic deliberation occurs best where citizens trust each other (Seligman

1992). Second, trust enables citizens to associate together so as to achieve objectives that they cannot reach individually (Gambetta 2000). Third, a trustful political atmosphere prevents disputes from escalating into social conflicts that would be difficult and costly to resolve (Sztompka 1996).

The importance given to interpersonal and social trust for the functioning of democratic societies explains why the term has become one of the key words in comparative political science and political sociology (Levi and Stoker 2000). It is precisely the fear of fading social trust, and the consequent dissolution of the moral basis of Western political communities, that has given rise to the recent 'call for a return to civil society' (Seligman 1992, 11). This call, also implicit in much of the theorization of social capital, stresses the need to reassert a sense of shared communality that is constantly threatened by a growing self-interested individualism (Seligman 1992, 7).

At stake in the latter concern is not only the moral sustainability of Western polities but also the conditions for their economic development. As a key element in the development of effective, low-cost enforcement of contracts, trust is seen to foster efficiency in economic transactions of various kinds. Hence, according to Kenneth Arrow (1972, 357), 'it can be plausibly argued that much of the economic backwardness in the world can be explained by the lack of mutual confidence'. Echoing Arrow's conviction, Douglas North points to a similar causality in developing countries, arguing that a lack of social trust is 'the most important source of both historical stagnation and contemporary underdevelopment in the Third World' (North 1990, 54).

The conceptual intermingling between trust, civil society and national economy is well captured in the following excerpt from the classic *Principles of Political Economy* by John Stuart Mill (1848/2006, 131–2):

> Conjoint action is possible just in proportion as human beings can rely on each other. There are countries in Europe, of first-rate industrial capabilities, where the most serious impediment to conducting business concerns on a large scale, is the rarity of persons who are supposedly fit to be trusted with the receipt and expenditure of large sums of money.

Extrapolated onto the scale of broader society, interpersonal trust and the social cohesion that it reflects are seen to affect positively both economic performance and the quality of public government. With higher trust, economic transactions become more efficient, as there is less need to invest time, effort, and money in establishing written contracts or other kinds of guarantees for business transactions. Instead, the business parties can trust that their interests will not be violated by each other (Knack and Keefer 1997). The quality of public government, again, is enhanced by the more acknowledged sense of citizenship and the lively political participation that characterize high-trust societies. Checked by active citizens, politicians and bureaucrats are less prone to the sort of dishonesty that hinders the quality of policy making and lowers the efficiency of service provision in more corrupt public governments. Moreover, trust among the citizenry facilitates collective activity that aims at expressing preferences and exercising demands on government (Gambetta 1988; Tavits 2006).

A veritable magic circle seems to be at work in the relationships between trust, economic performance and good government. In a much cited article, Knack and Keefer (1997, 1252–3) go on to list at least eight beneficial consequences of high degree of trust in society. With reference to scholars such as Coleman (1988), Galor and Zeira (1993) and Putnam (1993), they argue that, first of all, individuals in high-trust societies spend less on efforts to protect themselves against unlawful violations of their property rights or exploitation in economic transactions. Second, entrepreneurs have more time to devote to innovation in new products or processes as they do not need to invest in monitoring their partners' behaviour. Third, businesses and governments are less dependent on formal institutions to enforce agreements. Fourth, politicians and bureaucrats are perceived as more trustworthy, making their policy pronouncements more credible. Fifth, this credibility allows individuals to make longer term investment decisions and technological choices when they know what to expect from the government and legislation. Sixth, by improving access to credit for the poor, trust fosters their enrollment in secondary education. Seventh, individuals may enjoy better performance of government institutions, including publicly provided education. Eighth, specialized education becomes more valuable to individuals when hiring decisions are influenced more by educational credentials than by the attributed status of applicants, such as through blood ties or personal contacts (Knack and Keefer 1997, 1252–3).

Importantly, and paradoxically, the beneficial consequences of trust for both the economy and the government are rooted in a particular kind of 'civil society': one composed of certain kinds of (non-)political subjects. In a critical essay on neo-Tocquevillian social theory, Michael Shapiro (1997) points out that the contemporary neoliberal discourses on politics tend to point to the past as a model of civic life, thus turning a blind eye on today's multiple forms, modalities and spaces of politics. In neglecting the present forces of social and political containment and resistance, as well as the different spatialities through which they take place, Putnam (1993; 1995) and his followers come to assume that today's civil society should still resemble the ideals depicted in Alexis de Tocqueville's treatment of the American democracy, as well as in the legacy of Western liberal political philosophy and praxis that his work helped to shape (Shapiro 1997; Miztal 2001, 373; Cannone in this volume).

From this follow three key observations concerning the ways in which the relationship between the concept of trust and geography turns paradoxical. First, the Putnamian interpretation of trust as a key component of a vibrant civil society derives largely from a nineteenth-century discourse that 'assumes a unitary social order from which citizen action is provoked, and restricts political action to influence on governmental decision makers' (Shapiro 1997). While clearly resonant with much of contemporary neoliberal politics and policy making, this understanding is based on a defunct concept of what counts as political and how politics is practiced in civil society (Miztal 2001). Second, rooted in a political philosophical discourse that has presented itself as the global model, the idea of trust as the glue that holds the civil society together also reproduces an antiquated understanding of the spatiality of politics. In Shapiro's (1997) words,

> a focus that treats the nation as a whole and presumes the existence of a unitary national society constitutes an aggressive non-recognition of the variety of incommensurate social spaces constitutive of the 'society' and the variety of different kinds of political enactment within different social venues.

The third paradoxical aspect in the geographies of trust expands on the previous two. The importance of social and interpersonal trust for national economic and governmental performance is prompted by a particular neo-Tocquevillean ideal of civil society that, in itself, is deeply rooted in the Western liberal political philosophy and replicated in the institutional politics of most Western countries. Owing to this, the concept derives much of its intellectual thrust and explanatory power from its position in the very apparatus through which Western political communities are accustomed to make sense of their conditions of existence, successes and failures.

Herein lies the double-hermeneutical moment of trust: the concept is constitutive of the phenomena it appears only to describe. Such an uncritically accepted hermeneutical circle becomes problematic at latest when ideals of civil society and democracy are forcefully imposed into contexts where their Western *modus operandi* is not found. The 'democratization of Iraq' is a case in point. What for many commentators appears as a 'nascent civil society' in Iraq (Rahimi 2008), may also be understood in terms of a long tradition of contestation between religious and political authorities, eschewing what the majority of the Western world holds as the key ideas of 'people' (*demos*) holding 'power' (*kratia*). The failure to see the 'cultural specificity of civil society' (Schwartz 2003, 2) accounts for many problems that Western countries have encountered in dealing with the non-Western political world.

The contemporary concept of civil society arises out of a specific historical experience of modernization in the West, but it does not remain a unique aspect of Western society as it can also be found in other cultural contexts. Bellin (1995, 121), for example, points out that the term 'has entered the discourse of the Arab world and become a central concept in current Arab debate over the direction of politics in the region'. However, what remains obscured by this almost self-evident relevance is the fact that it is painfully difficult to specify where the analytical edge of the concept lies, and whether trust is actually a cause or consequence of the social and political phenomena it is taken to account for (Welch et al. 2005). Coupled with the assumption that the geographical shape of modern societies can be adequately defined by nation-state territories, the generic position of trust has acquired a hidden geographical dimension that nevertheless informs much concrete theoretical and empirical analysis of trust as a social phenomenon. It is this geography of trust that we shall turn to next.

Taking Space for Granted: The Hidden Geographies of Trust

The philosophical and theoretical complexity pertaining to the concept of trust has not rendered concrete analyses and empirical investigations unfeasible. On the contrary, the recent upsurge in research on social capital has also given a boost to empirical studies seeking to chart the degree to which people trust, for example, politicians, the government, societal institutions, foreigners and each other. As a result, an extensive literature has emerged providing evidence of the degree of social trust varying from place to place. Typically these studies are based on analyses of data that has been collected experimentally by means of questionnaires or structured interviews inquiring about the degree to which the respondents say they are willing to trust various targets (for example, nationalities, ethnic groups, professions, officials, neighbourhoods, etc. – see Rahn et al. 2003). An increasing number of studies are using survey data collected through representative samples assessing the climate of trust that exists in specific societies or communities (Delhey and Newton 2003). Most empirical studies have made use of data collected on an individual level and then aggregated for the purpose of analysis to the level of community, locality, region or nation (Welch et al. 2005).

While the use of experimental and survey data has been criticized by some scholars on account of measurement errors (see Miller and Mitamura 2003), or the poor validity of data caused by sampling errors or response bias (Knack 2000), it has been pointed out by others that the results nevertheless seem to correspond well with outcomes that can be expected on the basis of other data on the social groups and communities in question (Glaeser et al. 2000; Delhey and Newton 2003; Rahn et al. 2003). Hence an extensive literature now exists indicating differences in the amount of social trust that appear between nations, within regions of one country or between communities in one sub-state region or locally (see Robinson and Jackson 2001; Delhey 2007).

Comparative data from cross-national survey projects have yielded results according to which the level of social trust is higher in certain regions of the world, such as Scandinavia, and lower in other regions, such as Latin America (Knack and Keefer 1997; Inglehart 1990; see Siisiäinen in this volume). Empirical analyses have detected spatial differences at the sub-state level as well. Putnam's (1993) much cited study comparing northern and southern regions in Italy is a case in point. He later extended the analysis to the United States, with results indicating a pattern where Southern states show lower levels of social trust than the states in the Upper Midwest or in the Northeast (Putnam 2000).

While reflecting concern with geographical variation in the amount of interpersonal or social trust, there are major problems in this literature in terms of the manner in which trust is assumed to relate to space. The language of empirical analysis makes the geographies of trust appear quantifiable, showing the 'degree' or 'amount' of trust that exists within an aggregated data set representing a smaller or larger territory. A neighbourhood's, region's or nation's degree of trust may, thus, be read off from survey data measuring trust directly, or alternatively through variables such as 'associational activity' or 'value distance' used as shorthand for the existence of social trust (Delhey 2007).

Methodological critique aside (see Healy 2002), this approach to geography hides much more than it reveals. Territorial 'still images' that portray the social world in neatly demarcated geographical parcels, each characterized by differential levels or degrees of trust, end up misrepresenting the spatial dynamism of social processes and political landscapes. In a manner typical of quantitative analysis applying spatially aggregated variables, the key problem in these empirical studies derives from the representation of space as a fixed container for 'the social' – in this case social trust. Assessment of the relationship between trust and space is limited to regional or national differences in dependent or independent variables that are assumed to correlate with the measured amount of trust in a particular way. In other words, space is ultimately seen as local, regional or national level differences in social variables that are assumed to explain differences in the degree of social trust. As space, thus, merely 'contains' the social, the more dynamic and complex geographical ramifications of trust are never brought to the foreground, conceptualized or analysed.

The lack of any serious concern with the geography of trust has been noted by Welch et al. (2005, 468) who write that:

> Although the research to date has yielded a substantial body of findings about how individuals develop a sense of trust in those with whom they interact, and how trust affects other aspects of individuals' behavior, the research literature reveals comparatively little about the forces that shape trust, and are influenced by it, in group, organizational, and community settings.

They point out that one question especially deserving of researchers' attention is the 'radius of trust' issue first discussed by Fukuyama (1995) and Wuthnow (1998). Both authors argue that for Americans the radius of trust has progressively shrunk since World War II so that most Americans restrict their trust primarily to kin or friendship networks (Welch et al. 2005, 469). Yet, instead of attempting to account for the scalar retraction of the radius of trust, Welch et al. (ibid.) content themselves with pointing to 'the scale of human communities' as a potential source of explanation concerning 'conflicts that might derive from increasing diversity and the changing demographic composition of a community' with possible effects 'on the functioning of civil society and local institutions' (ibid., 469).

While issuing a notable invitation for further research, Welch et al. nevertheless fail to indicate where we could start looking for significant inroads into the geographies of trust. Indicative of the reasons for this omission, they conclude that the unanswered questions, 'all lead to situating trust within the broader contexts of community and society', but then elude the question of how the society or community 'should' be conceived of in geographical terms (ibid., 470). The theoretical direction they point to then is the taken-for-granted idea of 'society' as a socio-spatial unit defined by nation-state boundaries. Needless to say perhaps, this direction is but a dead-end street, a *cul-de-sac* of the mainstream modern social theory (see Häkli 2001).

The problem of hidden geographies of trust discussed above is intimately related to the implicit statism of modern social science thinking. The latter refers to the dominance of the state-territorial conception of 'society' in social science discourses, which have typically failed to address the question of what 'society' is in geographical terms (Agnew 1994; Taylor 1996; Häkli 2001). Unsurprisingly, this insensitivity to geographical assumptions about society is also present in literature dealing with the development of civil society in nineteenth-century Europe (Ehrenberg 1999). While an increasing reflexivity has emerged regarding the role of the state in the constitution of civil society, the geographical contexts through which the process is approached – cities, sub-state regions, nations or societies – are typically addressed in terms of a country defined and demarcated by the nation-state territory. This is what Taylor (1996, 1920) calls the 'embedded statism' of social scientific research.

Thus it is possible to talk about a 'hidden geographical agenda' in the study of trust as a key aspect of civil society, where the common assumption still is that the state territory adequately describes the spatiality of this 'society'. Even analyses that evince differing degrees of 'civility' between different parts of a country fall into this 'territorial trap' (Agnew 1994). For example, in arguing that the Southern Italian civil society is less supportive of efficient government than that of Northern Italy, Putnam (1993) operates on the assumption that the totality of the Italian nation-state forms a relevant societal context for comparison. Hence, what is 'outside' of Italy has little or nothing to do with the issue addressed purely in terms of the dominant geographical imagination of 'society.

The consequences of this hidden agenda include 'nationalization' of the study of trust; the indistinct use of the terms nation, country, and society; and the dominance of the country scale at the expense of more place-specific or transnational analyses of social trust (see Agnew 1993, 254; Wallerstein et al. 1996; Taylor 1996). Moreover, as discussed in the previous section dealing with the paradoxical geography of trust, the consequences of a state-centred concept of society can not be reduced to social theoretical discourse alone. Attention should also be paid to the myriad governmental practices that have produced and reproduced the nation-state as the dominant scale for representing the social world and framing issues of political importance (Häkli 2001). This social reality eventually came to be taken as a given not only in political philosophy and mainstream social research but also in numerous other spheres of social life, including governmental practices and the politics of 'domestic' non-governmental organizations and civic associations. This marked the consolidation of the double hermeneutic circle, linking together the importance of trust for political and economic viability of a 'society', and a nation-state as its territorial container. Seligman (1992, 10–11) sees this as one of the paradoxes of modern society: social trust is represented by the universalization of citizenship as the foundation of moral agency in the public arena secured by the state.

It is this co-construction of the civil society and the 'state-society' that causes the search for the geographies of trust to remind us of a dog chasing its tail. Because of the hidden statism in the understanding of civil society, norms of social trust are already framed in country terms deeply embedded in the politics of modern governmental and non-governmental institutions. This embeddedness is based on a shared geographical imagination through which the social world is enacted: the conception of space as a container for social relations, and 'society' as a territorially confined unit defined by the national state.

Little wonder then that even though social trust has been viewed as a phenomenon pertaining to communities of various sizes and compositions, the spatial ramifications of trust remain a neglected topic. Empirical studies have come to take administrative regions as their starting point, due to availability of statistical data or the set up of experimental sampling. This holds true also for country-level analyses as well as cross-country comparisons: the state territorial conception of civil society, defined by national boundaries, is largely taken for granted (see also Brewer et al. 2004).

The persistence of hidden geographies of trust is particularly intriguing in studies where political spatial units make little sense in understanding the phenomena in question. Environmental issues are a case in point. But even International Relations research on 'international' trust is typically based on representative national samples, thus operating on the basis of a state-centred concept of society (Brewer et al. 2005; Tennberg 2007). However, challenges to 'nationalized' understanding of trust have started to arise on the basis of recent geopolitical changes, such as transnational networking and new forms of cross-border governance, both of which have foregrounded emergent new forms of political space that evade the traditional communities of morality, politics, involvement and participation (Kramsch and Hooper 2004).

Indeed, the contemporary interdependent world with local and transnational processes intertwining and interacting, is calling into question identities, practices and discourses that were constructed in the context of the modern state-centred world (Castells 1997; Ley 2004; Staeheli 2008). New geographical imaginations about society as networks, flows and fluid spaces have started to emerge, undermining the domination of the Euclidean conception of space as container, and giving rise to spatial conceptions and identities that differ from those of the previous generations (Mol and Law 1994; Law 2002; Häkli 2008). People engage in trusting relationships in the context of local and global interconnectedness that is further reinforced by geographical imaginations produced and reproduced through 'contentious politics' (Leitner et al. 2008) as well as by mobility, migration, the media and consumer culture (Blunt 2007).

These new identities and conceptions are negotiated in the flux of everyday life, but also in the context of state-based institutions and non-governmental organizations and activities. Hence, the enactment of social trust is a contested field where the modern state-centred view of the world meets emerging new identities, discourses and practices connected to, for example, transnational environmental concerns, multiculturalism or 'cosmopolitics' (Cheah and Robbins 1998; Pickles 2003). New configurations of civic consent and political participation – new geographies of trust – are being negotiated in the background of processes whereby different partly overlapping and partly differentiated 'civil societies' are simultaneously emerging.

The challenge of recognizing and unravelling the taken-for-granted geographies of trust is yet to be fully addressed. This volume is an attempt to pinpoint and explore some of the issues involved and take tentative steps toward understanding the complexity of the relationships between geography and the contested notions of civil society, trust and social capital. But in order to reach our first approximation of the given 'geographies of trust' we must still deal with the third reduction of geography in the analysis of social trust: the reification of space as the embodiment of 'trust culture'.

What Makes Society 'Civil': The Geographical Reification of Trust

Attempts to explain governmental and economic performance by social and cultural means have a long history in social thought. A body of knowledge that began as a ruminative discussion of political philosophy in works by authors such as Thomas Hobbes (1651/2006) and Alexis de Tocqueville (1835/2000) was taken to a different level in the groundbreaking analysis of Western political and economic development by Max Weber (1905/1985). His sociological mode of thinking has encouraged generations of scholars to study the role of civic culture and social life in the function of political and governmental institutions.

In the late 1980s and early 1990s this interest reappeared forcefully in the context of sociologically oriented political science as practiced by James Coleman (1988), and Robert Putnam (1993). While Coleman carried out his empirical work with a focus mainly on the United States, Putnam's study showcased Italy as an emblematic case for analysing the cultural embeddedness of political and governmental phenomena. Putnam, however, was not the first to take an analytical look at Italian civic culture. His work was preceded by more than four decades of political science experimentation using Italy as a geographical laboratory for testing theories and developing argumentation concerning political cultures, and later, geographies of (dis)trust.

The origins of this interest in the political life of Italy may be attributed to the country's peculiar political history and many internal divisions. It is also possible to speculate, however, about more personal reasons for this interest, tracing back to Charles Edward Merriam, an influential American political scientist who served as the American high commissioner of public information in Italy during World War I. In the words of Gabriel Almond (1996, 1) Merriam was the 'head of American propaganda efforts in Italy where he dealt with the lagging morale of that country in its conduct of the war … trying to prevent Italy from going the way of Russia'. This experience gave rise to Merriam's thought concerning the central role of what he termed 'civic education' in fostering civic solidarity. He developed a clear sense that the problem of nation-building stood at the heart of political democratization, and that various techniques of controlled dissemination of ideas and manipulation of popular symbols might be used effectively in the process (Oren 2000, 98).

Significantly, after the war Charles Merriam returned to the United States and took over the Political Science Department of University of Chicago, giving rise to a pioneering behavioural strand in political science (Almond 2004). Among the dozens of PhDs that this unit trained were such comparative politics scholars as Gabriel Almond, Sidney Verba and Lucian Pye, all of whom contributed to the placement of Italian society under a political scientific looking glass. Merriam's own research resulted in a volume called *The Making of Citizens* that sought to perform a comparative analysis of civic cohesion and civic training systems in France, Germany, England, Italy, Soviet Russia, Austria-Hungary, Switzerland, and the United States (Merriam 1931). At least partly inspired by the differing performances of the countries involved in World War I, it nevertheless failed to develop an overarching theory and was relatively soon forgotten (Almond 1996). However, it may well be Merriam's personal experience in Italy during World War I that spurred the extensive American scholarly interest in that country's political culture in the decades that followed.

The idea of civic associations as a key factor in local development was first addressed by political scientist Edward Banfield (1958) in his study *The Moral Basis of a Backward Society*, focusing on the political life of a small community in southern Italy. The study sought to explain the community's relative underdevelopment by going back to 'amoral familism', i.e. a dominant ethos favouring the maximizing of short-term material advantage for the nuclear family instead of furthering the interests of the group or community (Banfield 1958, 85).

In spite of an initially controversial reception, this book has served as an inspiration for much subsequent work, including the much cited research *Civic*

Culture: Political Attitudes and Democracy in Five Nations by Almond and Verba (1963) – an attempt to discern the key determinants of political equilibrium in properly functioning democracies. They found that interpersonal trust plays a central role in the formation of secondary associations, which in turn, are essential for democratic participation. Moreover, they concluded that:

> our own data will tend to support Banfield's claim that the Italian political culture contains unusually strong parochial, alienative subject, and alienative participant components [… evidencing a] widespread mood of rejection that affects the attitudes of the great majority of Italians toward their political system in all its aspects. (Almond and Verba 1963, 38)

Hence, the study showed that the level of interpersonal trust was relatively low among Italian respondents, indicting a prevailing condition of distrust in the country. Moreover, they argued that 'each kind of polity – traditional, authoritarian and democratic – has one form of culture that is congruent with its structure' (Almond and Verba 1963, 33). Thus there had once been established cultural patterns which possessed considerable autonomy and could influence subsequent political and economic development.

Echoing these observations, Joseph La Palombara (1965) interpreted statistics showing high degrees of dissatisfaction with the government in Italy to be an indication of widespread political alienation. He argued that one reason for this lay in the rigid system of social stratification in Italy. Low social mobility has formed class-based social barriers which, over several centuries, have become institutionalized, with an impact on Italian political culture that La Palombara (1965, 315) describes as follows:

> the stratification system aids and abets notions of extreme class conflict, contributes to a sense of futility that spills over into men's attitudes toward and evaluation of the political system, and helps to perpetuate patterns of leadership in the legislature and bureaucracy that are direct causes of disdain and hostility toward the political system.

La Palombara (1965) comes to the distressing conclusion that in the Italian society mistrust is so pervasive that it prevents any of the cooperative undertakings necessary for civilized existence.

Since the mid-1960s a broad Anglophone literature has emerged on the Italian political system (see Hennessy 1969; Graziano 1973; Morlino 1984; Woods 1992). Yet, in terms of impact on scientific discussions concerning the relationships between trust, civil society and political culture, Robert Putnam's (1993) study *Making Democracy Work: Civic Traditions in Modern Italy* stands out as a landmark. More than three decades after the classic text by Edward Banfield, Putnam and his associates Robert Leonardi and Raffaella Nanetti argued that democracies work better when supported by a long-standing tradition of civic engagement with public affairs (see Cannone in this volume).

The basic idea is that when individuals act together in local associations and other informal contexts, they create generalized trust as a by-product, thus generally improving the 'civicness' of the civil society. This is assumed to form an important public resource that influences not only the citizen's quality of life but also economic development, political cooperation and governmental performance. Using various indicators of policy implementation, Putnam showed that Northern Italian regions perform better than Southern regions, and found that these observations correlated with data on distrust and dissatisfaction reflecting the differing degrees of 'civility' in Italian regions (Putnam 1993, 63–83).

Putnam's study has been criticized for several omissions and weaknesses, including failure to distinguish between policy activism and administrative efficiency as aspects of government performance, and a myopic account of Italian political history (Sabetti 2000; Tavits 2006). Moreover, his conception of civil society has been found too voluntaristic and devoid of conflict, or in the words of Mouritsen (2003, 664), 'pacific, functionalist and self-propelling [growing] spontaneously where there is civic soil'. Thus, in vesting the idea of civil society with civic virtues such as trust, sense of duty, solidarity and public-spiritedness, Putnam comes to neglect the existence and value of conflict. In so doing he diverts from those conflictual aspects of civil society that may spur or constitute civility by teaching citizens to accept confrontation and compromise as the cement of the political condition (cf. Edwards and Foley 1998; Mouritsen 2003; Cannone in this volume).

The legacy of research into Italian political culture, as represented in the above studies, shares more than the idea of Italy as political culture characterized by some degree of alienation. Above all, it shows a strong tendency of relating trust to geography in a very particular way. Scholarly interest in the political development of Italian society, partly set in motion by American geopolitical concerns after the World War I, was progressively built up on the basis of an idea of 'territorially' confined political culture which, instead of illuminating the ways in which space is involved in political processes, merely ended up reifying it.

Reification as defined by Gordon Marshall (1998) in *A Dictionary of Sociology* refers to:

> The error of regarding an abstraction as a material thing, and attributing causal powers to it – in other words the fallacy of misplaced concreteness. An example would be treating a model or ideal type as if it were a description of a real individual or society.

In the case of representing Italy the abstraction vested with misplaced concreteness is political culture conceived of as a territorial geography of trust.

Starting with Merriam (1931), and probably before, research on the Italian civil society has related political culture to space in territorial terms, seeking to attain convenient shorthand for enduring national or sub-state regional differences in what Fukuyama (1995) has termed 'trust culture'. The constitution of these trust relationships is then conceptualized in terms of virtuous civic spaces defined as arenas of face-to-face interactions, such as voluntary associations (Foley and Edwards 1999).

While simplifying the overwhelming complexity of the social world for analytical purposes, this moment of reification comes at a price. It imposes social clarity, cultural transparency and spatial distinction upon socio-cultural processes and settings that are characterized by historical and scalar complexity. By foregrounding civil society as a regional entity defined by the nation-state 'Italy', or a sub-state spatial unit 'Southern Italy', this strand of comparative politics has ended up in a theoretical *cul-de-sac* where the dynamism of civil society as a social process and contested accomplishment is overshadowed by the idea of relatively stable and clear-cut geography of trust.

The reified geography of Italian (dis)trust has not only produced moribund political theory, but it has also shifted attention away from key questions, such as, 'What is the civil in civil society' (Mouritsen 2003, 650)? By representing civil society as a homogeneous field of social space glued together by generalized trust, the analysis has rendered invisible vital aspects in the multi-scalar constitution of civic engagement that could better be grasped in terms of relational or fluid spatiality. These include manifold active or emerging social movements that challenge established forms of political participation by engaging people in civic activism that escapes any pre-set territorial arenas (Edelman 2001; Leitner et al. 2008). This significant area of civic activism in itself underscores the fact that there exists no coherent civil society that could be isolated from class and interest-group differentiation and then thought of, in Putnamian manner, as disinterested towards contemporary socio-economic hierarchies and power geometries (Shapiro 1997; Mouritsen 2003).

Ramifications of the geographical reification of trust run wider than just simplified accounts of the multifaceted political cultures in Italy. The understanding of civil society as a territorially circumscribed cradle for civility, where 'feelings and ideas are renewed, the heart enlarged and the understanding developed only by the reciprocal action of men one upon another' (Tocqueville 1835/2000 cited in Putnam 1993, 90), makes the concept fundamentally apolitical by associating it with civilized manners. However, as Mouritsen (2003, 664) aptly points out, civil societies are created and may be lost.

> We should certainly reflect on the concept at a time when globalization and localization challenge the format of nation states. Such forces require us to concentrate on the cool, political and conflict-accepting character of civil society … And we need ways to maintain broad political identities which treat nations as neither

> 'functioning' systems nor vessels of culture. Putnam's intervention is a symptom of the fact that civil societies are fragile accomplishments.

Instead of reifying trust as a spatially circumscribed resource for the common good, we should seek theoretical inroads into understanding the specifically situated historical organization of social and political space that provides people with uneven opportunities to enact and practice their civic engagements. We should re-think trust as an aspect of pervasive and differentiated civic engagement that arises, not through virtuous behaviour, but in connection with the multi-scalar and complex politics of everyday life.

Conclusion

In this chapter I have explored some of the ways in which geography has figured into attempts in the social sciences to understand and conceptualize trust-related phenomena. This exploration led to the unveiling of three distinct, yet interrelated, geographies of trust: paradoxical, hidden and reified. What these geographies have in common is their taken for granted nature, that is, received understanding concerning space as the container of 'the social', coupled with assumptions about the manner in which trust is related to spatiality. It is these implicit geographical aspects of trust that I have sought to unravel in this chapter.

My first take on the geographies of trust led me to suggest that any attempt to locate or pinpoint trust in modern societies is somewhat *paradoxical* because the concept is part of the very means by which modern polities have been constituted and by which we make sense of these same societies. Trust is deeply embedded in more than two centuries of modern political philosophy, underlying democratic institutions and civil society, informing both the ways in which governmental institutions function in relation to the 'sphere' of civil society and the ways in which their achievements are assessed. It is precisely the latter aspiration that has given rise to the flood of empirical studies charting the amount or degree of trust that people are willing to invest either in their political institutions or the citizens of their neighbouring countries (e.g. World Value Surveys).

While apparently paying attention to the explicit geographical variation in trust-related phenomena, this research interest has, in reality, worked to *hide* rather than illuminate the geographies of trust. Even though social trust has been viewed as a phenomenon pertaining to communities of various sizes and compositions, empirical studies have taken administrative regions as their starting point, thus assuming that space is best understood as the container of differential amounts of trust. The state territorial conception of civil society, defined by international boundaries, is predominantly taken for granted also in country-level analyses and international cross-country comparisons. The spatial assumptions that prevail in empirical studies of trust have not been adequately addressed in this research strand, and therefore, what seems an assessment of geographical variation merely works to conceal the spatial complexity pertaining to social trust.

I termed the third taken-for-granted spatiality of trust a *reified* geography. The term refers to the manner in which some geographical settings have been made showcase examples and emblematic cases of (dis)trust. I illustrated this reification by discussing the long legacy of reductionist uses of 'Italian society', and particularly its North/South division, in arguing for a particular Tocqueville-inspired understanding of the relationship between trust, civil society and economic or governmental performance. Arguing against the appealing simplicity of this theoretical model, I share the critical concern according to which such a reductionist reading of Italian society is highly problematic (Cannone in this volume).

In this chapter I have not been able to go much beyond a tentative exploration of the three given geographies of trust, but I have discussed some aspects of the development of contemporary societies that further undermine the relevance of paradoxical, hidden and reified understandings of geographies of trust. I argue that the dominance of the nation-state as the frame for political change and cultural signification is being eroded by the growing interdependence with local and transnational processes intertwining and interacting. Along with authors such as Shapiro (1997) and Mouritsen (2003), I maintain that this development is gradually calling into question identities, practices and discourses that were constructed in the context of the modern state-centred world order. Evincing this, new geographical configurations and imaginations about society as networks, flows and fluid spaces have started to emerge, calling into question 'the lazy certainties of the geometrical intuitions' (Bachelard 1994, 220). Consequently, social trust is now being enacted in the context of new figurations of contentious politics embedded in geographies of trust that defy simple measurable Euclidean spatiality.

To grasp these challenging geographical configurations of trust it is necessary to re-think the notion of civil society. In this chapter I have argued that civil society should not be conceived of as 'being' as much as 'becoming'. In other words, civil society should not be understood as a circumscribed entity that can be socio-spatially isolated from the market and the state for analytical purposes. Rather it is best understood as a multi-layered and multi-scalar social 'accomplishment' that relates to different spatialities in highly complex ways. In terms of topological spatialities discussed by Mol and Law (1994) and Law (2002), we can conceive of civil society and the relationships of trust it involves as co-constituted through different interconnected yet unconformable spatialities, such as scaled, regional, networked and fluid spaces (Leitner et al. 2008). These spatial types are not to be understood as mutually exclusive ontologies, but rather as abstractions deriving from different modalities that pertain to various kinds of institutional and individual action, including the agency of the 'more-than-human world' (Whatmore 2006, 602).

Indeed, if we take seriously the idea of civil society as a site of political contention rather than an oasis of civic consent, as I think we should, then we must strive to trace its 'becoming' in all possible political encounters between agents, locales, institutions, objects and events – whatever shape and content these may take. It is these 'multivalent and co-implicated spatialities of contentious politics' that Leitner et al. (2008, 158) propose as the challenging object for the study of contentious politics that 'challenges dominant systems of authority, in order to promote and enact alternative imaginaries'. This formulation clearly parallels the idea of civil society presented here, even though their use of the concept points to a regrettably categorical understanding of civil society as distinguishable from 'the state' and 'the market' (Leitner et al. 2008, 157). To push forward the analysis of the geographies of trust it is imperative to open up the concept of civil society and explore its co-constitution with hegemony-challenging contentious politics. It is equally crucial to ask what role, if any, relationships of individual and social trust play in these processes, and how the agents of contentious politics tactically perform and strategically bend different unconformable, yet interrelated spatialities.

Bibliography

Agnew, J. (1994), 'The territorial trap: the geographical assumptions of international relations theory', *Review of International Political Economy* 1:1, 53–80.

Almond, G.A. (1996), 'The civic culture: prehistory, retrospect, and prospect', ,<http://repositories.cdlib.org/csd/96-01>, accessed 10 January 2008.

Almond, G.A (2004), 'Who lost the Chicago School of political science?', *Perspectives* 2:1, 91–3.

Almond, G.A. and Verba, S. (1963), *The Civic Culture: Political Attitudes and Democracy in Five Nations* (Princeton NJ: Princeton University Press).

Arrow, K. (1972), 'Gifts and exchanges', *Philosophy and Public Affairs* 1:4, 343–62.

Bachelard, G. (1994), *Poetics of Space.* Translated by M. Jolas (Boston: Beacon Press).

Banfield, E.C. (1958), *The Moral Basis of a Backward Society* (Glencoe, IL: The Free Press).

Bellin, E. (1995), 'Civil society in formation: Tunisia', in A.R. Norton (ed.).

Blunt, A. (2007), 'Cultural geographies of migration: mobility, transnationality and diaspora', *Progress in Human Geography* 31:5, 684–94.

Brewer, P. R., Aday, S. and Gross, K. (2005), 'Do Americans trust other nations? A panel study', *Social Science Quarterly* 86:1, 36–51.

Brewer, P. R., Gross, K., Aday, S. and Willnat, L. (2004), 'International trust and public opinion about world affairs', *American Journal of Political Science* 48:1, 93–109.

Castells, M. (1997), *The Power of Identity.* Vol. 2 of *The Information Age: Economy, Society and Culture* (Cambridge: Blackwell).
Cheah, P. and Robbins, B. (1998), *Cosmopolitics: Thinking and Feeling beyond the Nation* (Minneapolis: University of Minnesota Press).
Coleman, J. (1988), 'Social capital in the creation of human capital', *American Journal of Sociology* 94, 95–120.
Cox, K., Low, M. and Robinson, J. (eds) (2008), *The SAGE Handbook of Political Geography* (London: Sage).
Delhey, J. (2007), 'Do enlargements make the European Union less cohesive? An analysis of trust between EU nationalities', *Journal of Common Market Studies* 45:2, 253–79.
Delhey, J. and Newton, K. (2003), 'Who trusts? The origins of social trust in seven societies', *European Societies* 5:2, 93–137.
Edelman, M. (2001), 'Social movements: changing paradigms and forms of politics', *Annual Review of Anthropology* 30, 285–317.
Edwards, B. and Foley, M.W. (1998), 'Civil society and social capital beyond Putnam', *American Behavioral Scientist* 42:1, 124–39.
Ehrenberg, J. (1999), *Civil Society: The Critical History of an Idea* (New York and London: New York University Press).
Etzioni, A. (1993), *The Spirit of Community: Rights, Responsibilities, and the Communitarian Agenda* (New York: Crown Publishers).
Ferguson, A. (1767/1980), *An Essay on the History of Civil Society* (New Brunswick, NJ and London: Transaction Publishers).
Foley, M.W. and Edwards, B. (1999), 'Is it time to disinvest in social capital?', *Journal of Public Policy* 19:2, 141–73.
Fukuyama, F. (1995), *Trust: The Social Virtues and the Creation of Prosperity* (New York: The Free Press).
Galor, O. and Zeira, J. (1993), 'Income distribution and macroeconomics', *Review of Economic Studies* 60:1, 35–52.
Gambetta, D. (ed.) (1988), *Trust: Making and Breaking Cooperative Relations* (Oxford: Blackwell).
Gambetta, D. (2000), 'Can we trust trust?', in D. Gambetta (ed.).
Giddens, A. (1976), *New Rules of Sociological Method* (London: Hutchinson).
Glaeser, E.L., Laibson D., Scheinkman J.A. and Soutter C.L. (2000), 'Measuring trust', *The Quarterly Journal of Economics* 115:3, 811–46.
Gramsci, A. (1929–35/1988), *An Antonio Gramsci Reader* (New York: Schocken Books).
Graziano, L. (1973), 'Patron-client relationships in Southern Italy', *European Journal of Political Research* 1:1, 3–34.
Häkli, J. (2001), 'In the territory of knowledge: state-centered discourses and the construction of society', *Progress in Human Geography* 25:3, 403–22.
Häkli, J. (2003), 'To discipline or not to discipline, is that the question?', *Political Geography* 22:6, 657–61.

Häkli, J. (2008), 'Regions, networks and fluidity in the Finnish nation-state', *National Identities* 10:1, 5–22.

Healy, T. (2002), 'The measurement of social capital at international level', paper prepared for International Conference on the Measurement of Social Capital ,<http://www.oecd.org/dataoecd/1/60/2380281.pdf>, accessed 6 February 2008.

Hegel, G.W.F. (1821/1991), *Elements of the Philosophy of Right* (Cambridge: Cambridge University Press).

Hennessey, T.M. (1969), 'Democratic attitudinal configurations among Italian youth', *Midwest Journal of Political Science* 13:2, 167–93.

Hobbes, T. (1651/2006), *Leviathan* (London: Continuum).

Hudson, A. (1998), 'Placing trust, trusting place: on the social construction of offshore financial centres', *Political Geography* 17:8, 915–37.

Inglehart, R. (1990), *Culture Shift in Advanced Industrial Society* (Princeton, NJ: Princeton University Press).

Knack, S. (2000), 'Trust, associational life and economic performance', paper prepared for the HRDC-OECD International Symposium on The Contribution of Investment in Human and Social Capital to Sustained Economic Growth and Well Being ,<http://www.oecd.org/dataoecd/6/31/1825662.pdf>, accessed 2 February 2008.

Knack, S. and Keefer, P. (1997), 'Does social capital have an economic payoff? A cross-country investigation', *Quarterly Journal of Economics* 112:4, 1251–88.

Kramsch, O. and Hooper, B. (eds) (2004), *Cross-Border Governance in the European Union* (London: Routledge).

La Palombara, J. (1965), 'Italy: fragmentation, isolation, and alienation', in L. Pye and S. Verba (eds).

Law, J. (2002), 'Objects and spaces', *Theory, Culture and Society* 19:5–6, 91–105.

Le Galès, P. and Lequesne, C. (eds) (1998), *Regions in Europe* (London: Routledge).

Leitner, H., Sheppard, E. and Sziarto, K.M. (2008), 'The spatialities of contentious politics', *Transactions of the Institute of British Geographers* 33:2, 157–72.

Levi, M. and Stoker, L. (2000), 'Political trust and trustworthiness', *Annual Review of Political Science* 3, 475–507.

Ley, D. (2004), 'Transnational spaces and everyday lives', *Transactions of the Institute of British Geographers* 29:2, 151–64.

Livingstone, D.N. (1995), 'The spaces of knowledge: contributions towards a historical geography of science', *Environment and Planning D: Society and Space* 13:1, 5–34.

Locke, J. (1690/1980), *Second Treatise of Government* (Indianapolis: Hackett Publishing).

Lorimer, H. (2003), 'Telling small stories: spaces of knowledge and the practice of geography', *Transactions of the Institute of British Geographers* 28:2, 197–217.
Marshall, G. (1998), 'Reification', in J. Scott and G. Marshall (eds).
Merriam, C.E. (1931), *The Making of Citizens: A Comparative Study of Methods of Civic Training* (Chicago: University of Chicago Press).
Mill, J.S. (1848/2006). *Principles of Political Economy, with some of their Applications to Social Philosophy* (Oxford: Oxford University Press).
Miller, A.S. and Mitamura, T. (2003), 'Are surveys on trust trustworthy?', *Social Psychology Quarterly* 66:1, 62–70.
Miztal, B.A. (2001), 'Trust and cooperation: the democratic public sphere', *Journal of Sociology* 37:4, 371–86.
Mohan, G. and Mohan, J. (2002), 'Placing social capital', *Progress in Human Geography* 26:2, 191–210.
Mol, A. and Law, J. (1994), 'Regions, networks and fluids: anaemia and social topology', *Social Studies of Science* 24:4, 641–71.
Morlino, L. (1984), 'The changing relationship between parties and society in Italy', *West European Politics* 7:4, 46–66.
Mouritsen, P. (2003), 'What's the civil in civil society? Robert Putnam, Italy and the republican tradition', *Political Studies* 51:4, 650–68.
North, D. (1990), *Institutions, Institutional Change and Economic Performance* (Cambridge: Cambridge University Press).
Norton, A.R. (ed.) (1995), *Civil Society in the Middle East*, vol. 1. (Leiden: E.J. Brill).
Oren, I. (2000), 'Uncritical portrayals of Fascist Italy and of Iberic-latin dictatorships in American political science', *Comparative Study of Society and History* 42:1, 87–118.
Parry, G. (1976), 'Trust, distrust and consensus', *British Journal of Political Science* 6:2, 129–42.
Pickles, J. (2003), *A History of Space* (London: Routledge).
Porta, P.L. (1998), 'Turning to Adam Smith', *Atlantic Economic Journal* 26:1, 102–15.
Putnam, R.D. (1993), *Making Democracy Work: Civic Traditions in Modern Italy* (Princeton, NJ: Princeton University Press).
Putnam, R.D. (1995), 'Bowling alone: America's declining social capital', *Journal of Democracy* 6:1, 65–78.
Putnam, R.D. (2000), *Bowling Alone.* (New York: Simon & Schuster).
Pye, L. and Verba, S. (eds) (1965), *Political culture and political development* (Princeton, NJ: Princeton University Press).
Rahimi, B. (2008), 'Ayatollah Ali al-Sistani and the democratization of post-Saddam Iraq', Nathan Hale Foreign Policy Society Working Paper ,<http://www.foreignpolicysociety.org/workingpapers/WP6--Rahimi.pdf>, accessed 14 January 2008.

Rahn, W., Yoon, K.S. Garet, M. and Lipson, S. (2003), 'Geographies of Trust: Examining Variation in Generalized Social Trust Across Knight Communities', ,<http://knightfoundation.com/research/>, accessed 16 January 2008.
Rawls, J. (1971), *A Theory of Justice* (London: Oxford University Press).
Robinson, R.V. and Jackson, E. (2001), 'Is trust in others declining in America? An age-period-cohort analysis', *Social Science Research* 30:1, 117–45.
Sabetti, F. (2000), *The Search for Good Government: Understanding the Paradox of Italian Democracy* (Toronto: McGill-Queens University Press).
Schwartz, F. (2003), 'What *is* civil society', in F. Schwartz and S. Pharr (eds).
Schwartz, F. and Pharr, S. (eds) (2003), *The State of Civil Society in Japan* (Cambridge: Cambridge University Press).
Scott, J. and Marshall, G. (eds) (1998), *A Dictionary of Sociology* (Oxford: Oxford University Press).
Seligman, A. (1992), 'Trust and the meaning of civil society', *International Journal of Politics, Culture and Society* 6:1, 5–21.
Shapiro, M.J. (1997), 'Bowling blind: post liberal civil society and the worlds of NeoTocquevillean social theory', *Theory and Event* 1:1 ,<http://muse.jhu.edu/login?uri=/journals/theory_and_event/v001/1.1shapiro.html>, accessed 3 September 2007.
Sidaway, J. D. (2008), 'The geography of political geography', in K. Cox, M. Low and J. Robinson (eds).
Staeheli, L. (2008), 'Democracy and democratization', in K. Cox, M. Low and J. Robinson (eds).
Sztompka, P. (1996), 'Trust and emerging democracy: lessons from Poland', *International Sociology* 11:1, 37-62.
Tavits, M. (2006), 'Making democracy work more? Exploring the linkage between social capital and government performance', *Political Research Quarterly* 59:2, 211–25.
Taylor, P.J. (1996), 'Embedded statism and the social sciences: opening up to new spaces', *Environment and Planning A* 28:11, 1917–28.
Tennberg, M. (2007), 'Trust in international environmental cooperation in Northwestern Russia', *Cooperation and Conflict: Journal of the Nordic International Studies Association* 42:3, 321–35.
Tocqueville, A. de (1835/2000), *Democracy in America.* Translated by G. Lawrence (New York: Perennial Classics).
Wallerstein, I. et al. (Gulbenkian Commission) (1996), *Open the Social Sciences: Report of the Gulbenkian Commission on the Restructuring of the Social Sciences* (Stanford, CA: Stanford University Press).
Weber, M. (1905/1985), *The Protestant Ethic and the Spirit of Capitalism.* Translated by T. Parsons (London: Unwin).
Welch, M.R., Rivera, R.E.N., Conway, B.P., Yonkoski, J., Lupton, P.M. and Giancola, R. (2005), 'The determinants and consequences of social trust', *Sociological Inquiry* 75:4, 453–73.

Welch, M.R., Sikkink, D. and Loveland, M.T. (2007), 'The radius of trust: religion, social embeddedness and trust in strangers', *Social Forces* 86:1, 23–46.
Whatmore, S. (2006), 'Materialist returns: Practising cultural geography in and for a more-than-human world', *Cultural Geographies* 13:4, 600–609.
Woods, D. (1992), 'The centre no longer holds: The rise of regional leagues in Italian politics', *West European Politics* 15:2, 56–76.
Wuthnow, R. (1998), *Loose Connections: Joining Together in America's Fragmented Communities* (Cambridge, MA: Harvard University Press).

Chapter 2
Searching for Social Capital

Mauro Cannone

> Our civil life is suffering in America today. Citizens are working together less and shouting at each other more. The common bonds of community which have been the great strength of our country from its very beginning are badly frayed … We see our families and our communities all over this country coming apart … It's hard for a lot of overworked parents to find the time and space for those things that strengthen the bonds of trust and cooperation … If you go back to the beginning of this country, the great strength of America, as de Tocqueville pointed out, … has always been our ability to associate … and to work together to find common ground. And in this day, everybody has a responsibility to do more of that. If we work hard – and work together – … we can lift America's fortunes once again.
>
> President Clinton, State of the Union Address 1995

Introduction: Conceptualizing the 'Social' Side of the Economy

During the 1990s, one of the main intellectual endeavours of a significant number of social scientists was the attempt to address more rigorously the old idea that the mechanisms of economic life are not only governed by the exact laws uncovered by economic analyses, but they are also influenced by the wider social fabrics into which these laws are woven. In particular, this multidisciplinary reflection aimed to uncover the 'capital' value of 'social' relations, and the conceptual focus of this increasingly influential research stream is now more generally known by the name of 'social capital'. Unsurprisingly, moreover, the initial attempts to conceptualize these immaterial aspects proved to be especially attractive for mainstream economists, who have always been somewhat intrigued by the existence of some 'irrational' or 'informal' aspects complementing their rational explanations of market mechanisms. With their crucial support, this search for a more precise definition of such 'mysterious' human aspects was enthusiastically underpinned by national governments, international institutions, and non-governmental organizations, and eventually adopted as a potentially innovative explanation for conditions of both 'development' and 'underdevelopment', and thus as a new conceptual basis for policy-making initiatives in many fields (Harriss and de Rienzo 1997).

According to Bagnasco (1999), the success of such academic attempts is usually announced by a significant change in the terminology employed to define the issues they address, and this change normally contains much more than a simple introduction of new linguistic styles or conventions. Rather, it often indicates a major conceptual shift in the ways in which analytical questions are acknowledged and discussed. In this case the change was also announced by a particularly symbolic episode attesting to the triumphal entry of the concept of social capital into the political arena: in 1995 US President Bill Clinton invited Robert Putnam, a political science professor from the University of Michigan, to Camp David. The topic of their discussion was the decline of associational life and the spread of individualism within American society. Since that meeting, Putnam's name has gained fame far beyond the academic walls to which it had been confined until then, and so has the concept of 'social capital' discussed on that occasion. Within this concept, Putnam gathered together a set of social and cultural conditions that, in his view, underpin or hinder political effectiveness and economic development (see Field 2003).

President Clinton was certainly fascinated with the analysis of the 'social illness' affecting his country, an illness that Putnam identified first in an article (1995) and then expanded in a book (2000) with the metaphorical title of *Bowling Alone: America's Declining Social Capital.* According to the findings of these works, the familiar image of 'bowling teams' – fellows gathering together to play, chat, and drink – was no longer representative of contemporary America, as staying at home and watching TV alone had slowly become the main leisure activity of individuals, especially those living in alienating urban environments. But besides his fascination with such research in general, President Clinton was probably much more concerned with its conclusions, according to which these rising levels of individualism are strongly jeopardizing the very basis of the democratic and economic system in what, to him, was the paradigmatic expression of democracy and (free market) economy. In other words, Clinton was alarmed because the 'stock' of social capital on which the US had built its fortunes was, according to Putnam, seriously at risk.

Used as an independent variable to explain not only political performances but also the economic success of a number of localities at various scales of analysis, social capital was envisaged in Putnam's research as a by-product of the 'culture' of these localities, namely as a historical predisposition towards cooperation and association stemming from specific civic attitudes rooted in some places but not in others (see Häkli in this volume). In providing a mono-causal explanation for a broad range of socio-economic phenomena such as the political and economic effectiveness of countries, regions and cities, this concept eventually established itself within the lexicon of many financial and political institutions, especially in the West, generating a wide interest in studies, statistics and policy areas of different disciplines. In influential academic journals and key political think-tanks the concept of social capital was repeatedly invoked as a panacea, reflected in words such as 'associations', 'community', 'cooperation' and 'trust', as in President Clinton's *State of the Union Address* of 1995 (see initial quotation).

However, as I mentioned above recalling Bagnasco's observations, the vocabulary employed to define specific ideas is often a key tool in emphasizing some particular aspects and nuances of these same ideas, while inevitably overshadowing others. Therefore, the choice of some words cannot be simplistically envisaged as an irrelevant semantic detail, and for this fundamental reason in the following sections I will seek to provide an explanation for the choice of those words very often associated with, or even taken as synonyms of social capital by Putnam and his followers. In other words, I will attempt to show that his work on the concept became much more than a mere academic exercise. In fact, by drawing attention (back) to the fundamental externalities embedded in social life,[1] the results of his study seemed to match perfectly with the urgent need to employ similar conceptual conclusions for political and economic purposes. To uncover what has become a dominant use of Putnam's research findings, I will first trace the recent 'historical' contours that led to his successful (re)discovery of the economic significance of sociability, and to the successive popularization of some words used to describe its beneficial effects instead of others. Secondly, I will discuss the theoretical implications entailed by such words, as well as the crucial political and theoretical issues that they can – too easily – obscure from view.

Theories of Social Capital

As several scholars have noted (see Field 2003; Woolcock 1998; Woolcock and Narayan 2000), the basic ideas lying at the heart of the concept of social capital have an immediate appeal, for they coincide with the old, common assumption that 'it is not what you know, but whom you know' that counts. As this popular saying intuitively summarizes, by means of our social relations we are able to work together to accomplish tasks that we either could not attain by ourselves, or that we could achieve only with great efforts. What is new in the contemporary focus on this old idea, however, is the intention of uncovering the ways in which our relationships – encompassing familiar ties, friendships,

1 As Hirsch (1976) remarked long ago, the economic externalities concealed in social life have in fact been constantly marginalized by neoclassic economic theory, despite the original lessons of the masters of liberal thought such as Adam Smith, John Stuart Mill, and Alfred Marshall, to name but a few. And as a result of their marginal position, these externalities have always been kept safely locked within the allegedly peripheral deposits of knowledge of the social sciences. The concept of social capital was a crucial tool to (re)shift the interests of (neo)liberal economists towards the consequences of sociability, but the problem, as I will argue, is that their attention was drawn almost exclusively to James Coleman's and especially Robert Putnam's research, which employed a conceptual language and a methodological approach that did not entail a deep reconsideration of the economic orthodoxy, but instead complemented its ideological positions with fundamental sociological support.

memberships in groups, etc. (the 'social') – constitute a valuable resource (the 'capital') to rely upon when we are in need of help.

To be sure, the importance of relationships of association and mutual aid has been explored by an endless number of seminal thinkers and scholars, from Aristotle to Tocqueville, Durkheim and Marx, to recall but a few. Yet interest in this vast and fundamental topic had never truly given rise to a thorough systematic investigation of the matter, nor has it been included within a broader framework to conceptualize and define more rigorously how relationships can be revalorized, mainly for social and economic purposes (Portes 1998). When the term 'social capital' was first introduced in sociological disciplines by Lydia Hanifan in 1916, and when it reappeared in Jane Jacobs' *Life and Death of Great American Cities* (1961), a rather weak and provisional conceptualization was still associated with it. In fact, it was only in the 1980s and then especially in the 1990s that the (economic) values concealed in social relations began to be explored in more depth, and social capital progressively became a much more widely discussed topic:[2] one that 'brings a new focus and introduces new questions;' one which 'draws attention to the links between the micro-level of individual experiences and everyday activity, and the meso-level of institutions, associations and community' (Field 2003, 7). Put simply, the recent attention towards the concept seems to conceal an attempt at rephrasing, in academic and then political terms, the conditions under which the old assumption that 'relationships matter' can explain a range of 'mysterious' outcomes.

As a result of this recent success, the term 'social capital' has gone from being under-investigated to suffering from semantic inflation (Ventrone 2003), with too many definitions and uses that risk blurring its heuristic potential. In fact, the original ideas and intuitions implied by the concept gave rise to a sudden wave of multidisciplinary attention to it before a systematic and precise definition could be formulated. And this is precisely what made social capital a multifaceted and controversial concept, employed in a variety of fields with a wide range of different meanings, and often with contrasting overtones. As a consequence, the risk now associated with it is that of 'trying to explain too much with too little [and of] being adopted indiscriminately, adapted uncritically, and applied imprecisely' (Lynch et al. 2000, 404), while an agreement on its precise meaning is still missing (Nahapiet and Goshal 1998; Sabatini 2003; Sobel 2002).[3]

In other words, the rapid upsurge of interest in social capital studies has gradually produced a growing theoretical confusion over the very definition of

2 As a recent study points out, 'before 1981, the number of journal articles listing social capital as a key word totalled 20, between 1991 and 1995 this has risen to 109, and between 1996 and March 1999 the total was 1003' (National Statistics 2001, 6).

3 Sobel (2002, 144), for instance, reports the case of an edited volume where, of a total of 13 articles, 'at least nine contain extended discussion of what social capital means'.

the concept. I argue, one of the most significant reasons for such confusion, is that these studies have drawn varyingly from the three original conceptualizations which can be recognized as central in recent debates on social capital: those of Pierre Bourdieu, James Coleman, and Robert Putnam. These scholars have generally been credited with its introduction into theoretical discussion and its translation from popular metaphor into academic concept (Field 2003). As I will make clear in the following sections, much of the contemporary bewilderment stems precisely from the ways in which the nuclear idea envisaging social relations as capital has passed from one to another of these scholars, maintaining the same name and the same basic baggage of metaphorical assumptions, but from crucially different theoretical and political perspectives.

Pierre Bourdieu

With no apparent relation to the first, provisional appearances of the term in the works of Hanifan and Jacobs, social capital was (re)introduced in the field of social sciences by Pierre Bourdieu, who employed the general notion of 'capital' to account for the class divisions of contemporary French society, and to uncover the dynamics through which these divisions are reproduced. Bourdieu initially theorized the notion of 'cultural capital' – which was to become a fundamental conceptual pillar of his works – in one of his first books, *The Reproduction* (Bourdieu and Passeron 1974), where it was employed to explain the mechanisms by which cultural titles and qualifications gain social significance. Eventually, the notion of capital was extended to two other fields of analysis, or 'forms of capital': the economic and the social. With a Marxist approach, Bourdieu envisaged 'capital' as a 'stock' of tangible and intangible resources which altogether foster the hierarchical social structure of capitalism, reproducing social differentiation and exclusion. In his analysis, material wealth (economic capital), 'education' and 'taste' (cultural capital) and networks and contacts (social capital) are resources first inherited from the family and then accumulated through 'labour'. Crucially, they are also the tools employed by individuals to climb up the multilayered pyramid of Western societies. For example, those born in elite families will make use of their original advantages in terms of forms of capital inherited, and therefore their economic capital will likely be used to, say, access exclusive schools and universities where contacts and trust will be built with other privileged students (Bourdieu 1984). This and similar practical mechanisms explain the 'social stratification system' of Western democracies, and most importantly, the 'legitimization of dominant-class reproduction' (Adam and Roncevic 2003, 159), which in Bourdieu's view too often is erroneously taken as a natural, unquestioned process.

In particular, Bourdieu (1980) developed the concept of social capital to explain why individuals with similar backgrounds or equal stocks of economic and cultural capitals can reach different social positions. To him, the main reason lies in their different faculty of mobilizing the capital of a network of different acquaintances (that is, family members, old friends in elite schools, exclusive clubs, and so on), where a lesser or greater quantity of other capitals can be accessed. From these premises, Bourdieu (1986, 248) eventually provided a more precise and coherent definition of social capital as

> the aggregate of the actual or potential resources which are linked to possession of a durable network of more or less institutionalized relationships of mutual acquaintance and recognition – or in other words, to membership in a group – which provides each of its members with the backing of the collectively-owned capital.

Following this definition, the three main components of social capital are: (1) membership in a group, or the 'social'; (2) resources available to the group, or the 'capital'; (3) mutual (re)cognition, making potential resources effectively available to the groups' members. In other words, the size of social capital possessed by individuals has to be measured by the quantity and quality of connections they can effectively mobilize, and, in turn, on the quantity and quality of capitals potentially at the disposal of the network (Lin 2001; Portes 1998).

However, if social bonds of whatever type and capital in whatever guise constitute some sorts of 'tangible' elements, it is the more invisible role played by mutual cognition and recognition that switches the lights of social capital on, transforming it into a concept worthy of further investigation. Indeed, it is through cognition and recognition that social capital 'acquires a *symbolic character*, and is transformed into symbolic capital' (Siisiäinen 2000, 192, original emphasis) which is a crucial concept used by Bourdieu, firstly to clarify his theoretical analysis of capital(s), and secondly to show how everyday practices and agencies are linked to capitalist social structure. The unequal possession of economic, cultural and social capitals is a potential reason for class distinction, which becomes factual only when the differences in capital endowments are objectified by the work of mediation operated by the symbolic capital, whose role is indeed that of translating 'objective differences between groups or classes ... into symbolic differences and classifications that make possible symbolic recognition and distinction' (ibid.). For example, listening to classical music is considered a 'distinctive ability' in some social contexts, while in others it could be judged as a worthless waste of time: to Bourdieu (1984) the difference is concealed in the symbolic capital that this skill takes on in the different environments where individuals are set and their skills, objects, or relations recognized as more or less valuable.

The tools to operate this often spontaneous process of symbolic recognition are embedded in the *habitus* – another fundamental concept in Bourdieu's work – defined as 'a set of dispositions' and behavioural codes reflecting individuals' social positions explaining (essentially) the ways in which society (re)produces itself (Bourdieu 2000, 19). From childhood on, by acquiring the *habitus* of those who share the same physical and social space, people are predisposed to move, speak, think, behave in some peculiar ways, even and especially without conscious decisions because the *habitus* is physically and metaphorically embodied as a second nature that is created by, and simultaneously recreates practices of, distinction among social classes (Butler 1996; Cresswell 1999; Painter 2000; Valentine 2001). The *habitus* is then effectively converted into an operational practice through the concept of 'taste', or the acquired ability to judge by means of values emanated from the *habitus* itself (Bourdieu 1984). In other words the *habitus* shapes a system of classification, which manifests itself as an allegedly innate taste, acting as a compass for individuals in the open sea of society. For instance, lower social classes tend to choose their clothing with pragmatic and functionalist criteria, such as price and usefulness, and thus they follow the 'taste of the necessary' opposed to the aesthetic 'taste of gratuity and futility' usually privileged by the upper classes (ibid.).

Through cognition and recognition these same categories of *habitus* and taste set in motion the mechanisms of symbolic translation that take the social, and all other forms of potential capital, and turn them into 'meaningful and socially effective' (Siisiäinen 2000, 13) tools. In particular, the concepts of habitus, taste, and distinction have direct implications on social capital especially when Bourdieu's (1986) analysis, having explored 'how', focuses on 'where' it is made effective: social relationships do not happen in a void, but are rather the consequence of actors' 'proximity in physical (*geographical*) space or even in economic and social space' (ibid., 250, original emphasis) where their everyday contacts happen. In this space 'individuals as homogeneous as possible' (ibid., 251) are gathered by means of an endless production of 'occasions (… hunts, parties, receptions), places (selected schools, clubs, etc.), or practices (smart sports, parlor games, cultural ceremonies etc.)' where social exchanges occur, and where they create new relations while reinforcing the old ones. It is here, therefore, that cognition and recognition materially 'happen', and social capital is generated and converted eventually into economic and cultural forms by means of all these practices of reciprocity utilized by individuals that usually have peculiar characters in common, such as profession, income, gender, religion, ethnicity, or familiar background (Bourdieu 1986). As a result, those in a similar social position will likely share the same social and physical spaces with people of similar status, *habitus* and taste; and this at the same time is the consequence of and the mechanism (re)producing inequality in accessing resources.

More than as a predetermined asset which is socially or naturally inherited, Bourdieu thus envisages social capital as the result of a dynamic and relational process and as the consequence of an endless institutional endeavour aimed at creating and maintaining durable and useful relations able to secure material or symbolic advantages (Bourdieu 1986). With this seminal conceptualization, Bourdieu established social capital as a field of study in social research, from where it eventually spread to other disciplines, contexts, and countries. The concept emerged in the English-speaking world through the work of renowned American sociologist, James Coleman

James Coleman

In general, Coleman's work aims to explain the principles moving individuals to act in society (Field 2003), and in doing so he integrates the economists' principle of rational choice – namely the assumption that individuals act only to achieve their personal interests – with a more traditional analysis of social system, primarily focused on the role played by social norms, rules and obligations (Coleman 1988). While Marxist materialism was the clear basis of Bourdieu's analysis, Coleman was profoundly influenced by Nobel Prize-winning economist Gary Becker, one of the most eminent intellectuals of the Chicago School, who applied the theory of rational choice in his studies of human capital.

Coleman adopted this same approach to explore the origins of social organization, and in his theoretical construction he drew on the two main intellectual answers to the question of why individuals become social actors. While from a sociological perspective 'social norms, rules, and obligations' (Coleman 1988, 95) are the crucial factors leading to collective actions, for neoclassical economists like Becker social agency is a means of achieving personal interests. From these premises, Coleman constructs an all-encompassing theory in which he compensates for the conceptual flaws of a purely sociological approach by drawing on the strengths of economic principles. This aim was already evident in his studies on the roles of family, religion, and ethnicity in the differing educational achievements of pupils (Coleman et al. 1966; Coleman and Hoffer 1987; Coleman et al. 1982; Hoffer et al. 1985), and more particularly in his later monumental work, *Foundations of Social Theory* (1990).

Influenced by Bourdieu's early writings on the topic, in these and other later works, Coleman focuses quite extensively on the concept of social capital, but from a substantially different perspective. In his influential article 'Social capital in the creation of human capital' (1988), Coleman investigates the key roles played by the family and the community in creating different levels of 'human capital'– knowledge and skills developed by individuals mainly through education. More precisely, he shows how these levels are not simply the result of differences among educational institutions. Rather, they mirror different endowments of social capital embedded in the specific social environments by which students are surrounded. To Coleman, in fact, the positive action of social capital is concealed in a set of values, norms, and obligations that pupils – and individuals in general – feel compelled to follow. And since this set of ethical resources is particularly effective in what he defines as close, 'primordial' forms of social organizations – the church and the family – it is not a coincidence that students attending Catholic schools and living in more cohesive families and communities obtain better educational results. In fact, much more than those living in more 'secular' and open environments, they are driven towards higher performances precisely by the influence exercised by social capital, namely by the values and norms that the family and the surrounding community ask them to observe (Coleman 1988, 109–13). For Coleman, perceiving the need to live up to the expectations of those socially (and spatially) close to us – not to disappoint them – is a very strong personal incentive.

In other words, as Field (2003, 24) remarks, for Coleman social relationships 'constitute capital resources by helping to establish obligations and expectations between actors, building the trustworthiness of the social environment' and producing positive effects on various aspects of social life. For example, in his research on the diamond trade in New York, Coleman observes that the trustful relationships bonding buyers and sellers are very often inexplicable from a rational choice perspective, especially when merchants are allowed to privately examine samples of diamonds worth millions of dollars. The absolute faith that the samples will neither be stolen nor substituted is crucial to the functioning of this market, and such faith stems, in Coleman's view, from the fact that most of the traders of this micro-community are Jews, and are thus related by dense networks of relationships of many different kinds (for example, religion, ethnicity, neighbourhood, family). This environment, Coleman (1988) concludes, is endowed with large 'stocks' of social capital created by different types of relations. One of its consequences is that of lowering transaction costs and smoothing business agreements, and for this reason economic transactions flourish here.

As these examples show, social capital 'like other forms of capital ... is productive, making possible the achievement of certain ends that in its absence would not be possible'; however, in contrast with economic and cultural capital, it is 'less tangible ..., for it exists in the relations among persons' (ibid., 100) and moreover, following Coleman's definition, it is comprised of 'a variety of different entities, with two characteristics in common: they all consist of some aspect of a social structure, and they facilitate certain actions of actors – whether persons or corporate actors – within the structure' (ibid., 98).

Noticeably, whereas for Bourdieu social capital is a resource made available to individuals through social ties that open access to other, more tangible, forms of capital (economic and cultural mainly), for Coleman it inheres in social structures; it is a benign force generated as a by-product of social relations, which act, for example, as channels providing crucial information (regarding job opportunities, important events, etc.), or by forcing individuals to act according to certain norms and by providing effective sanctions that are ultimately beneficial to the whole community.

And since it is not possessed by nor does it belong to individuals, the social form of capital is a set of collective resources 'that inhere in family relations and in community social organization' (Coleman 1990, 300) such as voluntary associations, clubs, circles and similar structures where relationships[4] are particularly close. With these examples, so profoundly influential on Robert Putnam's successive works, Coleman (1988) defines the effects of associationism as always positive and beneficial, not only for the members of these associations but also for society in general, with no factual distinctions between the different aims and restrictions on access that tend to be the peculiar characteristics of social organizations. According to Coleman (ibid.), besides the accomplishment of their formal or informal objectives, the role of associations is also that of providing contexts and occasions to create and strengthen social relationships, in particular by offering to their members the opportunity to build mutual trust, exchange information and/or seek help on a number of different issues (such as employment opportunities and bank loans) that very often have little or nothing to do with the original aims of the associations themselves. Drawing on past case studies (see Lipset et al. 1956), Coleman also empirically shows how relations first established in associations can eventually become a form of 'capital' originating in spontaneous processes which have little to do with the purposes for which the associations were created. He argues that this capital value must be considered as a by-product which 'functions precisely because it arises mainly from activities intended for other purposes' (Schuller et al. 2000, 7) – a by-product of individuals' cooperative will and their mere act of 'being together'.

More generally, through the concept of social capital Coleman was able to frame some behavioural tendencies which rational choice theorists were not able to account for, and according to Fine (2002) this was just the first step in a wider process of the colonization of the social sciences by neoclassical economics, now provided with a concept – social capital – to clarify what economic models and formulas could not completely frame: human agency. This process, although set in motion by Coleman, rests on the fundamental contribution of Robert Putnam, who imported Coleman's approach to social capital into studies within his own discipline on the links between democracy, society and economics.

Robert Putnam

While Bourdieu and Coleman focused their analysis on the micro levels of social structure, with Putnam the concept stretched to the macro scales of regions and nations. In making this leap forward, however, the Putnamian version of social capital maintained Coleman's intellectual legacy while ignoring Bourdieu's

4 According to Coleman (1988), in more open social environments (for instance in large urban centres), relationships are generally fewer in quantity and also less effective in normative terms, because individuals do not feel the same constraining presence of a close community around them. For this reason, in such environments expectations, obligations, norms and sanctions have a much less normative function on social life.

conceptual distinction between resources and possessors. On this theoretical basis, social capital began its path towards success with the generic warning on the lowering levels of associationism in American society launched by *Bowling Alone* (1995; 2000). Notoriously, however, this research was stretching on the American ground the outcomes of an earlier investigation that Putnam conducted in Italy and that he published in his seminal work entitled *Making Democracy Work* (1993a). This study explains the different performances of formally identical political institutions, the Italian regional governments, as the result of the different level of 'civicness' embedded in their territories. Putnam (1993a) assumed that the high levels of efficiency of the northern and central regions and the stark, indolent attitude of public functionaries in the south are the result of socio-cultural factors. Namely, of different endowments of what Machiavelli called *vertu civile*, or civic virtues (see Putnam 1993a, 87) that are 'essential to stable and effective democratic institutions' (Putnam 1993a, 11). In his analysis, these different civic characters stem from particular 'cultural' traditions fostering or hindering mutual trust, cooperation, and the vibrancy of associational life that he measured with quantitative methodological tools.[5]

Such 'cultural explanation' will drive Putnam to demonstrate that the success or failure of government reforms in Italy are the consequence of history (and geography). In fact, 'the regions characterized by civic involvement in the late twentieth century are almost precisely the same regions where cooperatives and cultural associations and mutual aid societies were most abundant in the nineteenth century' (Putnam 1993a, 162). By contrast, the mistrustful and individualistic behavioural codes of southern Italians are envisaged as the present inheritances of past autocratic regimes of oppression and tight control exercised in these areas by several foreign dominations dating back several centuries. Because they are rooted in history and geography, these socio-cultural traditions are very difficult to erase. Moreover, besides influencing the present effectiveness of political and administrative organizations, they also determine the 'cultural' preconditions of economic life. And in fact, while the northern and central regions are among the most advanced in Europe, the south remains strikingly backward.

5 Putnam identified four key indicators to test the level of civicness of Italian regions: the vitality of social life (measured by membership in groups such as associations and organizations of different kinds), newspaper readership, electoral turnouts and participation to political life. All these indicators, according to the data gathered from broad national and regional surveys, showed a lower 'civicness' in the South than in the Northern and Central areas of the country. This realization brought him to conclude that societies blessed with civic attitudes achieve positive political performances, leading in turn to positive economic results. Interestingly, the first two indicators were taken directly from Tocqueville's definition of civic community, whereas the latter two have been employed in particular to measure the political activity of individuals across the country (Putnam 1993a).

As noted above, such observations constituted the theoretical and methodological foundation for Putnam's later studies in America, where he assessed the presence of what, after his Italian Grand Tour, he understood as the unique meter for revealing the social 'health' of a nation or a region: social capital. Like a doctor with a new magic formula in his luggage, Putnam travelled back to the US to practice the new techniques he learnt abroad, all consisting of a quantitative 'check-ups' on the vibrancy of civil society. The results were first published in the *Journal of Democracy*, and then eventually developed in a book (Putnam 2000) built around the same central assumption. In both of these works, Putnam documented the current decline in civic engagement in his country with the steady withdraw of citizens from the affairs of the community: 'the broader social significance' of the data gathered, Putnam (1995, 70) warns, 'lies in the social interactions and even occasionally civic conversations over beer and pizza that solo bowlers forgo'. And when individualistic behaviours undermine the quantity of social interactions and conversations as they do in contemporary US (Putnam 1995), the national 'reserve' of social capital diminishes dramatically, and as a consequence democracy and economic prosperity are seriously at risk.

Curiously, the constant erosion of social capital is described as an apparently mysterious phenomenon whose origin, in Putnam's analysis, does not lie in any of the major changes that transformed American society in the last 30 years or so. In other words, what Putnam (2002) defines as the 'strange disappearance of civic America' is neither the consequence of an increased residential mobility reducing local ties, nor the result of economic insecurity leading to extended working hours, nor the product of the massive movement of women into labour force (Putnam 1995; 1996; 2000). He concludes, rather, that all these trends have probably just emphasized the negative effects generated by the two main causes of the disengagement from any sort of social, and thus civic life: 'the technological transformation of leisure' and above all a 'generational shift' particularly evident by comparing the lowering number of association membership from the post-war years to the present (Putnam 2000; 2002). As no explicit cause has been identified for this change, Putnam recognizes the experience of war as a sort of national glue with positive outcomes towards collective life. Thus, restoring civic engagement in contemporary America, he suggests, 'would be eased by a palpable national crisis, like war or depression or natural disaster' (Putnam 2002, 1).

Underpinned by extensive data, statistics and surveys, this argument should be able to 'convince even a skeptic that something is going on and that there may even be a common cause of the trend' affecting American civicness (Sobel 2002, 140). And in fact, especially after the Camp David meeting, it convinced a large number of political and economic institutions worldwide to adopt the peculiar political and economic recipe that Putnam had 'discovered'. Namely, it convinced them to envisage social capital as a crucial resource concealed in the predisposition of individuals to associate, cooperate and trust each other and, more importantly, it convinced them to adopt this concept as a beneficial (and costless) principle of policy-making initiatives to promote 'development'.

The Theoretical Dichotomy of Social Capital

As it is manifestly apparent from this short historical overview of the origins of the concept, if the intuition that sociability can be considered as a form of capital opened a new Pandora's Box in social sciences and beyond, its conceptual definition has been addressed at least in two different ways, with Bourdieu drawing significant attention to its value as an individual tool of reproduction of economic inequalities, and Coleman and Putnam focusing, in the opposite direction, on its fundamental function to make almost any aspects of collective life – and thus also of the economy – work. Therefore, although both of these perspectives aimed at tracing the contours of a concept which falls under the same name, social capital, in effect they stem from strikingly different theoretical and political approaches.

What is even more remarkable, however, is that while Coleman's and especially Putnam's ideas were pulled out of the social capital box almost immediately, Bourdieu's work remained for some time largely forgotten at its bottom. In his 'sociology of conflict' (Siisiäinen 2000, 190), the analysis is in fact centred on individuals and their strategic ability to build and mobilize networks of relations with the aim of achieving some crucial goals. Profoundly rooted in the tradition of French Marxism, Bourdieu's key contribution resides in having uncovered the practical process through which invisible structural differences are translated into everyday practices and vice-versa, generating enduring effects of social reproduction. In particular, with his primary idea of society as a battlefield ridden with struggles Bourdieu, draws attention to the unequal possession of social relations, which 'can give individuals [unequal] access to crucial resources not otherwise available despite ample endowments of human or financial capital' (Edwards and Foley 2001, 10; see also National Statistics 2002). In doing so he was able to overcome the problem of how a structuralist account can be related to a constructivist approach to individual agency (Ritzer 1996). In his view in fact, the process of reproduction does not unfold unequivocally and deterministically from structure to practice, because it is mediated by the *habitus*, which in linking 'subjectivity to structure' (Schuller et al. 2000, 3) breaks the rigid Marxist mechanisms and 'becomes the structuring structure leading via practices to the development of new structures' (Siisiäinen 2000, 196).

Inevitably, some general theoretical questions have been raised in relation to the Marxist legacy of Bourdieu's critical sociology (see Latour 2005), and to some of the more particular flaws associated with it. For example, the fact that social capital is envisaged as a tool which can be legitimately deployed only in certain social environments has led to its use in criminal associations being largely overlooked. Likewise, Bourdieu's perspective regarding the effects of social capital as beneficial only for the elite seems to overlook situations in which even the least privileged individuals can effectively rely on their networks of relations to improve their social positions (Field 2003). However, as Schuller, Baron and Field (2000, 5) have observed, the lack of a deeper analysis encompassing these and other facets is perhaps related to the limited attention generally given to the concept, which, despite being one of the pillars of his social theory, 'remains curiously undeveloped' in Bourdieu's work compared to the weight and significance he accords to cultural and economic capital.

These weaknesses notwithstanding, Bourdieu's contribution is still 'arguably the most theoretically refined among those that introduced the term in contemporary sociological discourse' (Portes 1998, 3), mainly because he stresses that relationships can be considered as an asset only when they allow crucial resources to be controlled; whereas later, better recognized investigations, in envisaging relationships as inherently valuable, largely lose this distinction. Overlooking the dynamic and relational processes aimed at enacting these resources means neglecting by implication some significant issues, such as the original inequality of their distribution, the struggles over their use, and the relations of 'power' (*à la* Foucault) inevitably established between those who manage them and those who seek access to them.

Conversely to Bourdieu's approach, Coleman emphasizes the role of social structures, shifting the focus to the shape and the features of networks. In his analysis, social organization constitutes a capital resource through a rather vague process of social capital being translated into a set of different and often conflicting features, with the risk of it being envisaged more as an asset statically embedded in relations rather than as the result of a more dynamic process of construction and fruition of social networks. In Coleman's works, social capital lies in the existence of interpersonal links, and its influences are evident because of the positive outcomes that these links generate, on both personal and public levels. But for this reason, as Portes (1998) noted, by failing to conceptually and semantically distinguish 'what' is accessed from 'how' it is accessed, Coleman's definition of social capital becomes tautological, because envisaging 'social capital as equivalent with the resources ... obtained is tantamount to saying that the successful succeed' (ibid., 5).

Also, in contrast with Bourdieu, Coleman sees social capital as potentially beneficial for society as a whole (Baron et al. 2000; Field 2003), but he develops this view without addressing some crucial issues, such as those related to 'conflict, the elite and the systemic imperatives of (contemporary) capitalism', and this makes his, as well as Putnam's arguments, 'weaker and tamer' (Fine 2002, 796). Moreover, Coleman's (1990) veneration for 'primordial' forms of social organization such as the family and the Church demonstrates a rather conservative view of society, suggesting a nostalgia for a past in which the closed solidarity of the 'community', or the *Gesellschaft* in Tonnies' (1957) seminal definition, had not yet been replaced by the openness of a more artificial form of society (*Gemeinschaft*) which, in his view, inevitably confers to relationships less beneficial values. In such an over-optimistic evaluation of the closure of social bonds, however, Coleman fails to acknowledge 'the strength of weak ties', namely the fact that distant relationships are in fact 'indispensable to individuals' opportunities and integration' (Granovetter 1973, 1378).

From a broader perspective, Coleman aimed at uncovering how individuals' agency is linked and influenced by the social structure of which they are part. Social capital is used as a tool in this intellectual endeavour (Coleman 1988, 97). Coleman used this concept as a framework and then as a label for all those behavioural tendencies (such as mutual cooperation, for instance) that rational choice theorists saw as 'mysterious' exceptions to their rules. In his view they resulted from the 'invisible hand' of social norms essential to the more general functions of society and economy. In his analysis, following these norms is just a different way for social actors to pursue their own interests: a way in which their rational choices are concealed but equally present and active under the guises of solidarity or altruism. Similarly to Bourdieu's Marxist materialism, Coleman rejects *a priori* the possibility that people can cooperate, or even simply stay together for reasons other than a mere calculation of personal advantages (Field 2003).

This approach, so profoundly influential in shaping Robert Putnam's later conceptualization, emphasizes just the positive effects of social capital, and its value as a collective rather than private good, because it stresses the normative function that social capital plays at the individual or community level in regulating the wider mechanisms of society as a whole. From the same assumptions, Putnam will unsurprisingly draw his idyllic picture of social capital as an attribute of urban, regional or national communities where individuals work together in happiness and harmony as their particular interests converge and then benefit the whole society. Resembling somewhat Thomas Moore's Utopia, in the 'prosperous communities' (Putnam 1993b) of contemporary northern and central Italy and post-war America a generalized civic culture makes democracy work and the economy flourish, as though here the open conflicts that usually characterize collective life had magically disappeared. More interestingly, they disappeared by virtue of a process in which civicness is spontaneously generated at the grass roots of society. But as Bagnasco (1999) has warned, such a contradictory account might alter social capital's explanatory power – a power that should rather bring to light the resources hidden in social relations. By contrast, following the way led by Putnam could mean falling into 'a discouraging sociology of good feelings or into an unrestrained ideology of development from below' (ibid., 119). Or, similarly, it could mean falling into a problematic 'sociology of integration' (Siisiäinen 2000, 190), namely a theoretical perspective which, by taking a large picture of society, utterly overlooks what Foucault (1979) notoriously defined as the 'micro-physics' of the power that moulds every social relation.

Put simply, with its quantitative and positivist methodology, this sociological approach envisages society as being comprised of individuals treated as mere statistical numbers, and as a result it by-passes all the conflicting and problematic issues that inevitably shape human interactions. As Fine (2002, 797) observed, in this way it draws a conclusion based on 'false empirical observations' precisely because, as I shall explain, it loses sight of individuals – who become just passive products of the 'time' and 'space' in which they live – and of the complex dynamics influencing their social behaviours. For this reason, such empirical observations need to be analytically scrutinized so as to prevent them from giving rise 'to theory that subsequently takes on a life on its own' (ibid.). In other words, Putnam's sociology conceals some potentially dangerous assumptions to be identified and rigorously addressed before triggering politically conservative interpretations that could go beyond his original intentions (DeFilippis 20042).

Exploring the 'Social Capital of Integration'

As a number of authors have observed, particularly in regard to Putnam's use of the concept of 'civil society' (Berman 2001; DeFilippis 2004; Edwards and Foley 2001; Whittington 2001), these problematic assumptions stem paradoxically from a fundamental misreading of Tocqueville, whose work has explicitly inspired this American political scientist. More precisely, they stem from an approach to the notion of 'civil society' that tends, purposely or not, 'to emphasize those aspects of the concept that best suit … particular ideological or theoretical purposes' (Edwards and Foley 2001, 4).

For these authors, in fact, Putnam's approach is particularly representative of a recent neo-Tocquevillian tradition that has completely overlooked Antonio Gramsci's intellectual contribution. According to the Italian scholar (1975), the fundamental role of 'civil society' is that of denouncing the weaknesses embedded in and the inequalities generated by the political and economic system in which we live. This critical scrutiny could, therefore, strive to expose the weaknesses and inequalities present in the neoliberal system (of government). But in dulling Gramsci's critical edge through a process of selecting some aspects of his reflections while overlooking other crucial parts, neo-Tocquevillians defined as 'civil' merely that part of society that acts 'virtuously' (the voluntary sector), or does not at least engage in conflictual activities promoting political and economic changes of any sort. In this way they re-narrate the concept as the necessary precondition for well-functioning democracy, rather than a means of questioning or challenging it in any way (Edwards and Foley 2001; see also Couch 1968).

Therefore, by imbuing social capital with ethical and moral significance, neo-Tocquevillians in general and Putnam in particular envisage associationism and its consequences as always beneficial to both individuals and society. But as Levi (1996) and other scholars (Edwards and Foley 1998; Tarrow 1996) have remarked, the practical mechanisms and also the 'political' conditions under which membership in organizations and networks lead to a more effective democracy remain obscure; likewise the alleged influence that groups such as bird-watching societies and soccer clubs (see Putnam 2000) can impinge on economic mechanisms (Mohan and Mohan 2002). Furthermore, the choice of measuring individuals' participation in collective life not only erases the non-quantifiable dynamics and meanings involved in participation; it also carries some methodological issues that raise several concerns (Adam and Roncevic 2003; Foley et al. 2001; National Statistics 2001). In fact, Putnam 'seems to be unable to deal with theoretical or conceptual arguments that cannot be measured and analyzed quantitatively' (DeFilippis 2002, 792), and the scale of his 'grand measurements' is defined by aggregating survey responses from communities circumscribed by the borders of regions or nations.

The very idea of administrative political units as homogenous spatial containers filled by similar traditions and behaviours is, I would argue, the conceptual consequence of a strikingly problematic geographical imagination. In this view, the practical processes that – much more than 'cultural' and 'historical' similarities – have informed the contemporary organization of regional or national structures, have been completely overlooked. In Italy for example, soon after the proclamation of the Republic in 1946 the division of the national territory into twenty regions followed the proposal of an economist to adopt the same boundaries drawn one century earlier with the empirical purpose of gathering census data (see Vallega 1983). At that time, regions were generally identified as 'territorial organisms' (see Vidal de la Blache 1921) sharing similar physical landscapes, economic features, and moral and civic traditions. These similarities however were so generic and subjective that they could have easily been applied to justify any other kind of administrative division (Vallega 1983). The choice made by the Italian Constituent Assembly in 1947 thus clearly privileged an unproblematic approach to the epistemological question of what, in fact, is a region, and of how should it be determined; geographers' answers to these questions are by no means as simple and straightforward as the ones prompted by political bodies, as it happened it Italy and elsewhere (see Della Dora and Minca forthcoming).

Similarly, in Putnam's works the presence of common cultural traits enclosed within exact regional and national borders is taken for granted, resembling very closely the conceptual implications of a specific geographical tradition first established in the US by Carl Sauer (1925; 1941; 1956) and then reinforced by his student Wilbur Zelinsky (1973). Notoriously, Sauer focused on the ways in which human 'cultures', working with and on nature, create different landscapes endowed with distinctive personalities, which he terms as 'culture areas'. In this geographical imagination 'culture' is the key actor reified on the materiality of the landscape, but as Baker (2003) and Mitchell (2000) have critically observed, in the Sauerian tradition 'culture' permeates a particular locality by simply gaining passive acceptance from the group of people living in that locality, through a reproductive process ambiguously reinforced by history. Duncan (1980) also remarks that Sauer used the concept in a totalizing and mono-causal fashion, engendering a form of 'cultural determinism' that – I argue – was particularly evident in the successive works of Zelinsky (1973), who envisaged 'culture' as a 'superorganic' entity with which he explained a wide range of different phenomena.

The legacy that this influential tradition left not only in the post-war US, but also in contemporary academic geographies, as well as in their more popular ramifications in the English-speaking world in general (for a more detailed reconstruction of this enduring success, see Minca 2007), seems not only to have influenced Putnam's ideas, but also to have laid the groundwork for the broad and uncritical acceptance that they eventually received across the Atlantic, especially in political organizations and economic institutions. Remarkably, Putnam in fact sees social capital as a civic attitude rooted in a 'culture' which is geographically (pre)determined because it is inscribed within tangible spatial containers akin to the Sauerian 'culture areas', and thus it is also reproduced historically in some regions or nations (central and northern Italy; post-war America) but not in others (southern Italy or the contemporary US). Similarly to Sauer's and Zelinsky's accounts, in this understanding 'culture' is treated as a pre-existing, reified entity used to explain social behaviours at different scales of analysis. But as Mitchell (2000) argued, in geography as in other academic disciplines, in its vernacular perception as well as in more attentive theoretical constructions, every time the word 'culture' is employed, the questions one should be asking is: what indeed is culture? What is it made of? Where does it come from, and does it explain or rather is it 'to be explained'? These are the same ontological questions that Putnam systematically fails to address in his Italian and American journeys, unless explaining 'culture' with 'history', 'traditions' and 'behaviours' is not to be regarded as a tautological, but rather as a theoretically sound answer.

Crucially, it is as a consequence of such geographical imagination and of the methodological choices it entails that social capital has made a further leap forward in its short history, where the analysis has moved remorselessly from the individual to the community and ultimately to the regional and national levels of society (DeFilippis 2002). This final analytical scale, however, is incommensurably problematic, as it not only obliterates the differences between localized social contexts (Foley et al. 2001), but most notably it presents a snapshot of society where real actors are completely ruled out by the broader social and economic structures in which, following these methods, they are envisaged just as passive elements (Bagnasco 1999). Based on this problematic assumption, those regions and nations where Putnam believes that 'civic virtue' is historically rooted have been depicted as 'a kind of paradise on earth, with citizens, state and economic networks intertwined in civilized harmony and mutual regard' (Amin 1996, 327). In this picture, post-war America appears as a conflict-free country in which riots, uprisings, demonstrations and similar movements 'against' the *status quo* are totally insignificant. And like post-war America, northern and central Italy are portrayed as one undifferentiated area in which a history of civicness explains not only the 'culture' of its citizens, but also the supposed absence of issues such as labour exploitation and political interventions which, in reality, played significantly in favor of their economic development (see Hadjimichalis 2006).

Moreover, as Arnaldo Bagnasco observed quite a long time ago in his seminal work on *The Three Italies* (1977), rather than the sum of two historically and culturally different territorial parts, this country is composed of three structurally interwoven areas: the First Italy of 'central economic activities' in the north-westerns regions; the Second Italy of 'marginal economy' in the south; and the Third Italy with 'peripheral functions' in central and north-eastern areas. In Bagnasco's view, the key to understanding the origins of the contemporary conditions in different parts of Italy is to envisage each of these as parts in relation to the others. Only in this way can the backwardness of the southern regions be explained historically – Bagnasco argues drawing on a clear Gramscian legacy – as the outcome of very particular political and social dynamics that led to the reproduction of the original divide. Only in this way, in fact, can the backwardness of these regions be recognized as a strategic condition to set and keep in motion the rapid industrialization of the First Italy, whose factories, for instance, could thus rely on a virtually unlimited reserve of labour necessary to strengthen their supremacy within the national borders, and to compete in international markets. To be sure, that which Banfield (1958) notoriously defined as 'amoral familism'[6] has always been a 'cultural' question deeply embedded in the history of the southern areas of the country, but the reproduction of this culture has also been fundamental to the preservation of the social and economic domination of both the industrial bourgeoisie of the north and the noble landowners of the south. After the reunification of Italy (1861), the strategic alliance between the national elites perpetrated this reproduction by means of very political tools, which Gramsci (1975; see also Graziani 1972) uncovered already at the beginning of the last century in the *clientelismo*, *assistenzialismo*, and *trasformismo*. Rather than 'social' and 'cultural' illnesses, as they are usually and simplistically understood, these were tools based on a subtle form of exchange, or blackmailing, in which the ruling class bartered for its electoral legitimization at the local level by offering some insubstantial improvements (underemployment instead of unemployment) to the living conditions of a large part of unprivileged population (Bagnasco

6 Banfield uses this term to describe an attitude privileging personal or 'familial' interests over disinterested and 'civic' forms of cooperation and mutual aid.

1977). The result, in fact, was to 'change things so that things stay as they are', as Tomasi di Lampedusa (1958) epitomized in one emblematic sentence.[7]

More recently, the same simplistic 'cultural' explanations have also been applied to account for the recent economic success of the Third Italy, which in a very similar fashion has been put in relation with 'history', albeit a history based on cooperation, trust and 'civicness' notwithstanding the fact that just a few decades ago these north-eastern and central regions were merely an extensive rural space laying in conditions of backwardness comparable to that of the south. But what these allegedly historical explanations seem to forget is that the very history of these areas has been characterized until recently by a peculiar form of economic and social organization: that of the *mezzadria*[8] (Bagnasco 1999; cf. Farinelli 2003), based on an almost infinite number of small and mid-sized farming activities scattered across the territory and run by nuclear families, with the help of their closest relatives and friends. It was precisely because of this pre-modern socio-economic structure that, as Becattini (1998, 29) ironically observed, 'we [Italians] did not know it yet, but we were already post-fordists… when elsewhere the fordist imperative was still ruling'. Almost untouched by modernity, familiar and kinship networks continued to embody the key social unit around which, in post-fordist times, a more advanced form of economy was being organized. To be sure, the role of local 'culture' in this development can not be overlooked, but it needs to be framed in association with the specific temporal and spatial circumstances in which it assumed its new significance (beside the seminal works of Bagnasco 1977 and Becattini 1998; 2000, see also Graziani 1998).

Once again, if on the one hand 'history' and 'culture' certainly did and do matter in Italy as in the US – as they do in every other country, region or place – on the other hand they still need to be framed along with all the mechanisms which have made them 'count' in the present. This is necessary in order to understand that their action is not sustained by inertial forces, which are very difficult to erase, but by specific interests and practices, which work to keep them alive. But from the very partial view of the 'Italian cultures' to the neo-Tocquevillian understanding of civil society – from the role played by political

7 In his novel *The Leopard*, taking place in the years of the Italian Reunification, this sentence ('Unless we ourselves take a hand now, they'll foist a republic on us. *If we want things to stay as they are, things will have to change*') was addressed to a Sicilian nobleman advising him to form an alliance with the Savoy who were reunifying the country, and to thus abandon his former alliance with the Kingdom of the Two Sicilies.

8 The *mezzadria* is a feudal form of sharecropping, originated and largely utilized in continental Europe during the Middle Ages. This traditional system consisted of the concession of a plot of land from a landlord to a *mezzadro* (tenant) and his family for cultivation, with the products and the profits to be equally divided between the two parties involved in such contract. In Italy, the *mezzadria* was abolished by law only in 1971, and it had a particularly deep and enduring impact on the organization of socio-economic relations (see Bagnasco 1999).

institutions to the principles of regional division – what is strikingly evident in Putnam's dangerous assumptions, as many scholars have observed (DeFilippis 2004; Farrel and Knight 2003; Fine 2004; Schuller et al. 2000; Siisiäinen 2000) is the complete neglect of any traces of conflict in society and of the ways in which power, in fact, shapes and conditions social bonds.

For this fundamental reason, as I have attempted to show, Putnam's investigation appears as a rather shallow report on how the democratic and economic performances of Italy and the US have been achieved – a report that, as a result of its search for a mono-causal explanation, has failed to frame the peculiar role of power and conflict not only between different interest groups, but also at the micro levels of society, within civic organizations and other such places where social capital is believed to take shape. The vertical dimension within these associations has been equally bypassed, despite Max Weber's (1911) famous observation that these 'places of cooperation' are also structured around relations of domination that follow clear hierarchical patterns.

In broader terms, by neglecting such issues, the mechanisms of modern capitalism remained unacknowledged (Fine 2002). Putnam, in fact, implicitly assumes that 'ethical', horizontal relationships nurtured in voluntary organizations make democracy and also the (free market) economy 'work for society as a whole', but in taking this assumption for granted he crucially leaves some fundamental questions unaddressed – namely, the question of why, in fact, the work of these organizations has now become particularly essential; and of how and in whose interest the neoliberal system (to which these organizations have become so essential) operates. However, as Fine (2002) suggests, it was precisely by 'forgetting' to address such key questions that the concept of social capital has found its place on a (political) pedestal.

Conclusions: Putting Social Capital on a (Political) Pedestal

The wide criticism that Putnam's idea of social capital has engendered, has by no means limited the popularity this concept has achieved in and beyond academic circles. To comprehend the reasons for this success, the 'Putnamian' understanding of social capital has to be placed within a broader and more complex historical and political framework. What is extremely problematic in Putnam's account of the decline of associational life in the US and Italy is the utter obliteration of the two most crucial phenomena taking place at the end of the twentieth century, namely the intertwined processes aimed at economic restructuring and the dismantling of the welfare state. In times characterized by 'globalized' pressure to cut public expenditure, a concept recognizing 'culture' as the reason for economic success is understandably a crucial theoretical tool, with two main intellectual implications, both aimed at the legitimization of the present neoliberal turn.

First and foremost, with such conceptual overtones, the use of social capital allows for what is known in the literature of social sciences as an exercise in 'blaming the victim' (DeFilippis 2002; Fine 2001; 2002; Ventrone 2003). This exercise has been magnificently performed in social capital studies by Francis Fukuyama (1995) who, drawing from Putnam's conclusions, compared the economic results achieved by various national capitalisms and explained their different outcomes in terms of trust (hence the title of his famous book), social capital, and thus 'culture'. For Fukuyama, in countries such as China, France and Italy the central role played by the family and the state in society and religion has prevented the development of more extended forms of cooperation and association. Conversely, in Japan, Germany and the US, individuals are much more devoted to setting aside their personal and familial interests and working together towards common goals. As a result, he argues, these countries lead the world's political and economic scenario. Fukuyama's message is enormously problematic but crystal clear: legitimated by a cultural background that creates large reserves of social capital, the most extreme (neo)liberal economies dominate in and beyond global markets, while those countries in less privileged conditions should blame their own 'culture' and should rather think of how they could increase their levels of trust and civicness (see Ventrone 2003).

By pointing his finger at 'culture' even more simplistically than Putnam, Fukuyama 'describes' a situation rather than 'explaining' its origins. And ironically, both authors indicate social capital as the sticking plaster that those individuals, regions, or countries in less dominant positions should stick on social wounds that they leave crucially unaddressed (Skocpol 1996). To simplify, it seems as though, with no regards to the contingent economic and social settings, in the type of ideal society Putnam and Fukuyama have in mind, citizens work together, are members of organizations, vote and take part in political life; but they do not protest, they do not complain, and more crucially, they do not ask why 'in fact' they should be more 'civic'. And then, when the social capital of this ideal society declines, responsibility falls on those same citizens, found guilty of not 'working together' while 'the economic and political changes wrought by the corporate and governmental establishment' (Portes 1998, 19) remain hidden as details of little or no significance in accounting for the reasons why they prefer watching TV rather than going out with friends. Even more ironically, thanks to such ideological use of social capital, these citizens are now asked to repair the nation's social fabric, woven by a corporate and governmental establishment that had left them behind to begin with (Skocpol 1996, 25). In other words, as Fine (2002) points out, placing social capital on a 'pedestal' has in fact served as a powerful theoretical tool to support and universalize the necessities and values of the dominant elites, so as to get the dominated to uncritically accept them. But the process of semantically camouflaging particular values, such as generalized trust or cooperation, into universalized values of 'civicness', should itself be subject to suspicion already on account of its supporting the diffusion of universal culture, which, as Bourdieu (1998) recalls, echoing Marx's and Gramsci's lessons, is a culture promoting the interests of those in power.

By means of very similar theoretical tools, this powerful discourse has been able to conceal the structural roots of the deep socio-economic changes of the last 50 years or so. During this period – concerning which Putnam bemoans a progressive decline in social capital in America even as it came to play an increasingly crucial role in the development of the Third Italy – post-Fordism was gradually beginning to replace the old Keynesian formula of the vertical organization of mass production and state intervention in the economy and society (see Harvey 1989). While these new patterns have imposed 'flexibility' as the keyword to follow in the process of restructuring that is usually referred to as globalization, the violent repercussions that they have brought on at both macro- and micro-organizational levels of society – such as unprecedented residential mobility, financial insecurity, an increase in working hours and so forth – have been studiously ignored by Putnam, Fukuyama and their followers. Such changes are now narrated by these contemporary 'social capitalists' (Fine 2002) not as the inevitable consequences of clear political choices, but rather as inevitable facets of inevitable global competition.

In other words, by applying the same subtle, neoliberal logic to their idea of social capital, Putnam and Fukuyama not only limit themselves to an uncritical description of these new conditions, but they also carefully avoid a deeper discussion concerning relations between these structural political changes and negative or positive trends in social capital. And while in Putnam's account politics is just a consequence of 'culture' and thus is completely removed from view, Fukuyama (1995) recognizes one crucial role that politics can play in the question of (under)development: that of wasting social capital (for a critical evaluation of Fukuyama's work, see Bagnasco 1999). Although it is not clear what politics should do, apart from promoting an unspecified 'togetherness', in his view it is very clear what governmental organizations should *not* do: address or correct social difficulties with tools such as the welfare state, which promote a passive dependence on the state by supporting the weakest, thus contributing to the erosion of their self-help networks (for example, by helping young mothers, it discourages the constitution of new families).

This concern leads to the second main implication inherent in the recent popularity of social capital. Whereas Fukuyama claims that the welfare state is an ominous political tool to be utterly erased from the political agenda because it can destroy such a vital resource, others, in a much more subtle way, present social capital as a costless proxy to replace the old and expensive welfare system. More accurately, these latter made use of the concept as a crucial pillar to support the institutions' and policy makers' need to justify and disguise the transfer of the burden of social welfare from the public to the voluntary sector – now glorified as the realm of civicness and cooperation.

Both of these specific understandings and uses of the concept have eventually given rise to a number of critical strands, some of which have highlighted, for example, the role of the welfare system as a basic condition for the creation of a generalized trust upon which social capital is based, especially in those countries where this system works efficiently (see Siisiäinen in this volume). In more general terms, however, a significant number of scholars have simply marked their intellectual distance from the theoretical weaknesses, and above all from the political implications, of Fukuyama's and Putnam's suggestions. These scholars emphasize in particular that while the conceptualization of the former manifestly aims to legitimize the most extreme neoliberal policies, the conclusions drawn by the latter are, if possible, even more problematic because they advance, in a very convincing and refined way, the idea that social capital is all we need to make democracy and the economy work.

But as one of these scholars has observed, conceived as such, social capital becomes the political Third Way, that is, *another* way to say with *other* words that '"there is no alternative" to the "there is no alternative" of neoliberalism' (Fine 2002, 799). And unsurprisingly, while the World Bank (1997) of the 'post-Washington consensus'[9] defined social capital as the 'missing link of development' (see Fine 1999; 2002; DeFilippis 2002; Ventrone 2003), in 1995 President Clinton clearly and loudly launched the same basic message to his fellow citizens (see opening quotation): there is no alternative to the fact that 'we' have to *work together* more, and shout at each other less; that 'we' have to find the time and space to *strengthen the bonds of trust and mutual aid*; that 'we' have a *responsibility* to do more to lift 'our' *fortunes* once again. (Un)surprisingly, however, he was less clear and loud in addressing the reasons why 'there is no *other* alternative' to 'our' contemporary social malaises, and perhaps he was rather imprecise in forgetting to specify, more clearly and loudly, whom, in fact, 'we' and 'you' are.

Bibliography

Adam, F. and Roncevic, B. (2003), 'Social capital: The recent debates and research trends', *Social Science Information* 2:46, 155–83.

Amaturo, E. (ed.) (2003), *Capitale sociale e classi dirigenti a Napoli* (Rome: Carocci).

Amin, A. (1996), 'Beyond associative democracy', *New Political Economy* 1:3, 309–33.

Bagnasco, A. (1977), *Tre Italie: la problematica territoriale dello sviluppo italiano.* (Bologna: II Mulino).

Bagnasco, A. (1999), *Tracce di comunità* (Bologna: Il Mulino).

Baker, A. (2003), *Geography and History* (Cambridge: Cambridge University Press).

9 According to Fine (1999), the 'post-Washington consensus' aims at erasing and replacing the previous Keynesian strategies, defined as the 'Washington consensus', followed by this organization.

Banfield, E. (1958), *The Moral Basis of a Backward Society* (New York: Free Press).
Baron, S., Field, J. and Schuller, T. (eds) (2000), *Social Capital: Critical Perspectives* (Oxford: Oxford University Press).
Becattini, G. (1998), *Distretti industriali e Made in Italy. Le basi socio-culturali del nostro sviluppo economico* (Torino: Bollati Boringhieri).
Becattini, G. (2000), *Dal Distretto Industriale allo Sviluppo Locale* (Torino: Borlati Borlinghieri).
Berman, S. (2001), 'Civil society and political institutionalization', in B. Edwards et al. (eds).
Bourdieu, P. (1980), 'Les capital social: notes provisoires', *Actes de la recherché en sciences socials* 31:1, 2–3.
Bourdieu, P. (1984), *Distinction: A Social Critique of the Judgement of Taste* (Cambridge, MA: Harvard University Press).
Bourdieu, P. (1986), 'The forms of capital', in J. Richardson (ed.).
Bourdieu, P. (1998), *Acts of Resistance: Against the Tyranny of the Market* (New York: New Press).
Bourdieu, P. (2000), *Pascalian Meditation* (Cambridge: Polity).
Bourdieu, P. and Passeron, J. (1974), *La riproduzione* (Rimini: Guaraldi).
Butler, J. (1996), 'Performativity's social magic', in T. Schatzky and W. Natter (eds).
Clinton, W. (1995), 'State of the Union address', reported in the *Washington Post*, on-line at <http://www.washingtonpost.com/wp-srv/politics/special/states/docs /sou95.htm>, accessed 22 January 2008.
Coleman, J. (1988), 'Social capital in the creation of human capital', *American Journal of Sociology* 94, 95–121.
Coleman, J. (1990), *Foundations of Social Theory* (Cambridge, MA: Belknap Press).
Coleman, J. and Hoffer, T. (1987), *Public and Private Schools: the Impact of Communities* (New York: Basic Books).
Coleman, J., Campbell, E.Q., Hobson, C.J., McPartland, J., Mood, A.M., Weinfeld, F.D. and York, R.L. (1966), *Equality of Educational Opportunity* (Washington DC: United States Government Printing Office).
Coleman, J., Hoffer, T. and Kilgore, S. (1982), *High School Achievement: Public, Catholic and Private Schools Compared* (New York: Basic Books).
Couch, J. (1968), 'Collective behavior: an examination of some stereotypes', *Social Problems* 15, 310–22.
Crang, M. and Thrift, N. (eds) (2000), *Thinking Space* (London and New York: Routledge).
Cresswell, T. (1999), 'Embodiment, power and the politics of mobility: The case of female tramps and hobos', *Transactions of the Institute of British Geographers* 24:2, 175–92.
DeFilippis, J. (2002), 'Symposium on social capital: An introduction', *Antipode* 34:4, 790–96.

Dell'Agnese, E. (ed.) (forthcoming), *10 parole chiave in geografia* (Turin: Utet)
Della Dora, V. and Minca, C. (forthcoming), 'Regione e potere: o, il potere della regione', in E. Dell'Agnese (ed.).
Duncan, J. (1980), 'The superorganic in American cultural geography', *Annals of the Association of American Geographers* 70:2, 181–98.
Edwards, B. and Foley, M. (1998), 'Civil society and social capital beyond Putnam', *American Behavioral Scientist* 42:1, 124–39.
Edwards, B. and Foley, M. (2001), 'Civil society and social capital: A primer', in B. Edwards et al. (eds).
Edwards, B., Foley, M.W. and Diani, M. (eds) (2001), *Beyond Tocqueville: Civil Society and the Social Capital Debate in Comparative Perspective* (Hanover, NH: University Press of New England).
Farrell, H. and Knight, J. (2003), 'Trust, institutions, and institutional change: Industrial districts and the social capital hypothesis', *Politics and Society* 31:4, 537–66.
Field, J. (2003), *Social Capital. Key Ideas* (London: Routledge).
Fine, B. (1999), 'The developmental state is dead – Long live social capital?', *Development and Change* 30:1, 1–19.
Fine, B. (2001), *Social Capital versus Social Theory* (London: Routledge).
Fine, B. (2002), 'They f**k you up, those social capitalists', *Antipode* 34:4, 796–9.
Foley, M.W., Edwards, B. and Diani, M. (2001), 'Social capital reconsidered', in B. Edwards et al. (eds).
Foucault, M. (1979), *Discipline and Punish: The Birth of the Prison* (Harmondsworth: Penguin).
Fukuyama, F. (1995), *Trust: The Social Virtues and the Creation of Prosperity* (New York: Free Press).
Gramsci, A. (1975), *Quaderni del carcere* (Torino: Einaudi).
Granovetter, M. (1973), 'The strength of weak ties', *American Journal of Sociology* 78:6, 1360–80.
Graziani, A. (1972), *L'economia italiana: 1945–1970* (Bologna: Il Mulino).
Graziani, A. (1975), *Crisi e ristrutturazione nell'economia italiana* (Torino: Einaudi).
Graziani, A. (1998), *Lo sviluppo dell'economia italiana. Dalla ricostruzione alla moneta europea* (Torino: Bollati Borlingheri).
Hadjimichalis, C. (2006), 'The end of Third Italy as we knew It?', *Antipode* 38:1, 82–106.
Hanifan, L. (1916), 'The rural school community center', *Annals of the American Academy of Political and Social Science* 67, 130–38.
Harriss, J. and de Renzio, P. (1997), '*Missing link* or analytical missing? The concept of social capital – an introductory bibliographic essay', *Journal of International Development* 9:7, 919–37.

Harvey, D. (1989), *The Condition of Postmodernity* (Cambridge: Basil Blackwell).
Hirsch, F. (1976), *The Social Limits to Growth* (London and Henley: Routledge and Kegan Paul).
Hoffer, T., Greeley, A.M. and Coleman, J.S. (1985), 'Achievement and growth in public and Catholic schools', *Sociology of Education* 58:2, 74–97.
Jacobs, J. (1961), *The Death and Life of Great American Cities* (New York: Vintage Books).
Latour, B. (2005), *Reassembling the Social: an Introduction to Actor-Network-Theory* (Oxford and New York: Oxford University Press).
Levi, M. (1996), 'Social and unsocial capital', *Politics and Society* 24:1, 45–55.
Lin, N. (2001), *Social Capital: A Theory of Social Structure* (Cambridge: Cambridge University Press).
Lipset, S., Trow, M. and Coleman, J.S. (1956), *Union Democracy: The Internal Politics of the International Typographical Union* (New York: Free Press).
Lynch, J., Smith, G.D., Kaplan, G.A. and House, J.S. (2000), 'Income inequality and mortality: Importance to health of individual income, psychosocial environment, or material conditions', *British Medical Journal* 320, 1200–204.
Minca, C. (2007), 'Humboldt's compromise: or, the forgotten geographies of landscape', *Progress in Human Geography* 31:2, 179–93.
Mitchell, D. (2000), *Cultural Geography: A Critical Introduction* (Oxford and Malden: Blackwell Publishers).
Mohan, G. and Mohan, J. (2002), 'Placing social capital', *Progress in Human Geography* 26:2, 191–210.
Nahapiet, J. and Goshal, S. (1998), 'Social capital, intellectual capital, and the organizational advantage', *Academy of Management Review* 23:2, 242–66.
National Statistics (2001), 'Social capital. A review of the literature', *Social Analysis for National Statistics*, Social Analysis and Reporting Division, Office for National Statistics.
Painter, J. (2000), 'Pierre Bourdieu', in M. Crang and N. Thrift (eds).
Portes, A. (1998), 'Social capital: Its origins and applications in modern sociology', *Annual Review of Sociology* 24:1, 1–24.
Putnam R. (1993a), *Making Democracy Work: Civic Traditions in Modern Italy* (Princeton, NJ: Princeton University Press).
Putnam, R. (1993b) 'The prosperous community: social capital and public life', *The American Prospect* 13, 1–8.
Putnam, R. (1995), 'Bowling alone: America's declining social capital', *Journal of Democracy* 6:1, 65–78
Putnam, R. (2000), *Bowling Alone: the Collapse and Revival of American Community* (New York: Simon and Schuster).

Putnam, R. (2002), 'The strange disappearance of civic America', *The American Prospect*, <http://www.prospect.org/cs/articles?article=the_strange_disappearance_of_civic_america>, accessed 29 January 2008.
Richardson, J. (ed.) (1986), *The Handbook of Theory: Research for the Sociology of Education* (New York: Greenwood Press).
Ritzer, G. (1996), *Sociological Theory* (New York: McGraw Hill).
Sabatini, F. (2003), 'Capitale sociale e sviluppo economico'. *Dottorato di ricerca in economia politica, serie working papers dei dottorandi*, No. 12, Luglio 2003.
Sauer, C. (1925), 'The morphology of landscape', *University of California Publications in Geography* 2:2, 19–53.
Sauer, C. (1941), 'Foreword to historical geography', *Annals of the Association of American Geographers* 31, 1–24.
Sauer, C. (1956), 'The agency of man on the earth', in L. Williams (ed.).
Schatzky, T. and Natter, W. (eds) (1996), *The social and political body* (New York: Guilford Press).
Schuller, T., Baron, S. and Field, J. (2000), 'Social capital: a review and critique', in S. Baron et al. (eds).
Siisiäinen, M. (2000), 'One concept, two approaches: Bourdieu and Putnam on social capital', *International Journal of Contemporary Sociology* 40:2, 183–202.
Skocpol, T. (1996), 'Unravelling from above', *American Prospect* 25, 20–25.
Sobel, J. (2002), 'Can we trust social capital?', *Journal of Economic Literature* 40:1, 139–54.
Tarrow, S. (1996), 'Social movements in contentious politics: A Review Article', *The American Political Science Review* 90:4, 874–83.
Tomasi di Lampedusa, G. (1958), *Il Gattopardo* (Milan: Feltrinelli).
Tönnies, F. (1957), *Community and Society* (East Lansing, MI: Michigan State University Press).
Valentine, G. (2001), *Social Geographies* (Harlow, Prentice Hall).
Vallega, A. (1983), *Regione e territorio* (Milan: Mursia).
Ventrone, O. (2003), 'Riflessioni sul concetto di capitale sociale', in E. Amaturo (ed.).
Vidal de la Blache, P. (1921), *Principes de géographie humaine* (Paris: Armand Colin).
Weber, M. (1911), *Geschäftsbericht*, in *Verhandlungen des ersten deutschen Soziologentages vom* 19.–22. Oktober (Tübingen), 39–61.
Whittington, K. (2001), 'Revisiting Tocqueville's America: Society, politics, and association in the nineteenth century', in B. Edwards et al. (eds).
Williams, L. (ed.) (1956), *Man's Role in Changing the Face of the Earth* (Chicago: University of Chicago Press).
Woolcock, M. (1998), 'Social capital and economic development: toward a theoretical synthesis and policy framework', *Theory and Society* 27:2, 151–208.

Woolcock, M. and Narayan, D. (2000), 'Social capital: implications for development theory, research, and policy', *World Bank Research Observer* 15:2, 225–50.
World Bank (1997), *World Development Report 1997* (New York: Oxford University Press).
Zelinsky, W. (1973), *The Cultural Geography of the United States* (Englewood Cliffs, NJ: Prentice Hall).

Chapter 3
The 'Magic and Loss' of Social Capital and Local Development

Paolo Giaccaria

Introduction

The purpose of this chapter is to consider the extent to which the notions of 'social capital' and 'local development' can be interpreted and discussed in conjunction with one another.[1] In particular, I argue that these concepts share particular characteristics, such that if we consider them together, we can usefully address certain questions about social capital and local development that remain unanswered in the literature. To date, the relationship between social capital and local development has not been investigated in depth, and yet the idea that social capital plays a significant role in local development has been given a great deal of credence. The role of social capital within local development tends to be discussed mainly in terms of the social dimension of economic processes. This emphasis on the social side of such economic processes is symptomatic of the 'cultural' and/or 'relational' turn in economic geography (Barnes 2001), which is typified by research examining the relationship between culture, institutions and economic development. The purpose of this chapter is not to review this complex body of literature in detail, but rather to focus on drawing out the connections between the cultural and relational turn in economic geography and ideas concerning social capital and local development.

Since the late 1990s, several different schools of thought have developed out of the cultural and relational turn in economic geography. In this broad branch of research, Amin (1989), for instance, undertook a critical analysis of Italian industrial districts; Californian economic geographers produced highly influential work on flexible specialization and transaction costs (Scott 1998;

1 In this chapter we use the term 'local development' instead of 'local and regional development' or 'local economic development', since important differences exist between those terms. In particular, the 'local and regional development' label implies a higher degree of continuity with post-war regional studies (Pike et al. 2007), while 'local development' is more commonly used to stress its discontinuity with the regional science tradition. The label 'local economic development' is instead normally more common in the policy-related literature, because of its focus on the economic dimension of local development compared to the social, the cultural and the political.

Scott and Storper 2003; Storper 1997); and the French Regulationist school conversely developed a Gramscian notion of Fordism (Boyer and Saillard 2001). Other noteworthy approaches are those developed by the international *Groupe de Recherche Europèen sur les Milieux Innovateurs* (Camagni and Maillat 2006), and by Porter with his notion of 'cluster' (Porter 1990), while Crang (1997) has been one of the first to explicitly provide an appraisal of the ways in which contemporary economic geography was rethinking the culture/economy nexus. The call for a more radical departure from mainstream economics was made even clearer in Ash Amin and Nigel Thrift's provocative editorial in *Antipode* (2000), and in the debate that it subsequently provoked. In the same period, other scholars preferred to employ the term 'relational turn' to mark an approach willing to keep economic geography linked to the regional and local development literature of the 1980s and the 1990s (see Bathelt and Glückler 2003; Yeung 2005). The debate has also been strongly influenced by the rise of the so-called *New Economic Geography* and by the related debate between geographers and economists (Martin and Sunley 1996; Martin 1999; Sheppard 2000a and 2001; Overman 2004; Marchionni 2004; Duranton and Rodriguez-Pose 2005).

Most of these approaches originally aimed at offering an emancipatory alternative to neoclassical economics by providing a new framework for understanding the relationships between cultures, economic development and place (Magnaghi 1990). These developing ideas however, quickly became popular amongst political and economic institutions, where they began to influence the discourses and practices of development on different scales, from the regional and urban governments to the European Commission (2005; 2007) and organizations such as the World Bank and the Organization for Economic Cooperation and Development (OECD 2002; 2007; World Bank 2004; 2006). The growing support for these ideas outside the academic sphere resulted in a complex framework of ideas and debates being translated into a generalized and supposedly universal set of policy prescriptions, such that it arguably brought about overly-simplified and crude understandings of, and approaches to, local development.

The fact that the same set of concepts had offered both a framework for emancipatory geographies and a basis for simplistic and superficial neoliberal understandings of the society/economy nexus largely explains why both the social capital and the local development debate in academia is now looking to backtrack (see Hadjimichalis and Hudson 2007). Often interpreted as the recognition of the social dimension of economic life, social capital and local development, in fact, became Trojan horses which helped to implement and strengthen neoliberal methodologies and individualistic accounts of society (Geschiere and Nyamnjoh 2001; Hadjimichalis 2006; Giaccaria and Scarpocchi 2008). Although originally envisaged as ways to undermine the domination of the economic over the social, the basic assumptions about concepts of social capital and local development were subverted, such that they actually became powerful tools for the promulgation of neoliberal economic interpretations of social and political life. This chapter thus reflects on this contradiction between the emancipatory aims of the cultural turn in economic geography and the actual outcomes of implementing these ideas in normative policy-making.

My main argument is that there is always something (methodologically) magical and (politically) lost when dealing with social capital and/or local development. Both of these concepts intrinsically rely on properties (such as trust, common values, institutional arrangements, shared public goods) that in some way connote the idea of the 'local' as a 'political enabler' for communities to attain their developmental goals. In discussions of these concepts, both the social capital and local development literatures assume that agency belongs not only to individuals, but also to a collective body. I would argue however, that this jump from the individual to the collective scale is often treated as a kind of *magic* passage. The magic here refers to the fact that – and it is my main proposition – no fully consistent 'micro-foundations' are given in order to explain to what extent, and under what conditions, agency transfers from the individual to the collective. This criticism has often been directed at Putnam's account of social capital (1993), especially in reference to his problematic belief that participation in associations directly translates into social capital. This shortcoming in the literature is a methodological one: both social capital and local development theories are grounded and legitimized on the basis of supposedly 'scientific' methodologies (such as methodological individualism). Such methodologies, however, are inconsistently applied within the social sciences, and more importantly, are also inherently political. This political dimension is apparent in the way in which the notion of organic collective agency allows discourses on social capital and development to be conflated with consensus-building narratives. Typically, such narratives rely on the belief of the existence of collective agency in order to present consensus as something that can be taken for granted, rather than as a product of negotiations between different shareholders.[2] Hence the magic in the methodology results in a loss in the political. This 'magic and loss' mechanism, or Foucauldian 'dispositif', is consequently something that critical approaches towards neoclassical and neoliberal economics must examine in greater depth.

The Magic and Loss of Social Capital and Local Development

Before introducing the 'magic and loss' dispositif into our analysis, we must briefly consider some of the features that are shared by the discourses on social capital and local development. I would suggest that the question is not

2 Despite its relevance, the importance of consensus-building in local and regional development has not been addressed in depth in literature. For a critical account of this issue, see for instance the recent article by Hadjimichalis and Hudson (2006) about the relationship between network literature, regional development and democracy.

simply whether social capital is a key asset in fostering local development or not (Mohan and Mohan 2002). Rather, our argument is based on the fact that a strong *theoretical and methodological proximity* exists between these two concepts. A first fundamental element of the proximity, or common ground, that exists between social capital and local development is that they have been conceptualized in similar ways: respectively, in terms of *collective agency* and *collective action.*[3] Despite the fact that these terms both sound similar and address the same methodological issue – that is, the micro-foundation of aggregate behaviours – these two concepts are not perfectly synonymous: the term 'collective action' is more often used in public choice literature,[4] whereas 'collective agency' is a term borrowed by the social sciences from cognitive studies (Swanson 1992; Bandura 2000; 2006; Castelfranchi 2005). The different implications of these two terms are important and should be recognized. Indeed, social capital, at least in Putnam's account, tends to be described as a form of 'public good', a concept traditionally linked with Olson's neoclassical account of collective action (Olson 1965). On the other hand, the literature on local development, especially in French and Italian, takes a more cognitive approach by conceiving of collective agency as an 'emergent property'[5] at the local scale (Conti and Giaccaria 2001; 2006). Frequent references in economic geography to such terms as 'system' (Cooke and Morgan 1998) and 'learning' (Maskell et al. 1998) are indicative of the influence of cognitive studies on this branch of the discipline.

I Want Some Magic to Sweep Me Away

Collective action and collective agency are therefore two fundamental theoretical and methodological reference points for both social capital and local development. This point is central to the present discussion and it deserves further reflection. Indeed, collective action and collective agency are essential in order to address the question of continuity and/or discontinuity between the 'individual' and the 'collective'. By making reference to collective action and to collective agency, both social capital and local development discourses are apparently

3 This in turn relates to what is probably one of the all-time primary issues of the social sciences, the 'structuration dilemma' – that is, the understanding of how human agency moves from individuals to society or community and vice versa. The literature reacting to Giddens's work (1986) on structuration is extensive and I do not have the space here to fully engage with such fundamental questions.

4 Following Hardin's early work (1982), the notion of collective action has been developed within public choice literature mainly by Ostrom over the last years (1990; 2000). See also Ostrom and Ahm (2003) on the relationship between the public choice literature and Putnam's approach to social capital.

5 In systemic analysis an 'emergent property' is a feature of the whole which the individual components of the system do not possess on their own and it is produced by interacting with other components.

well-rooted in strong micro-foundations: methodological individualism (Arrow 1994; Udehn 2001) and cognitive complexity, respectively (Byrne 1998). In this section, I argue that such foundations are rather unstable, grounded on some flimsy references to a metaphorical – rather than methodological – account of collective action and/or collective agency. This is the 'magic' side of our 'magic and loss' dispositif; the fact that the micro-foundation, the passage linking the individual to the collective, is never demonstrated explicitly, but just implied and thus left to the imagination. This results from the fact that the terms, and ideas of, collective action and agency tend to be understood as appealing metaphors, rather than rigorous methodologies.

The present approach thus engages with the numerous critiques of 'local development' and affiliated concepts that have emerged during the past couple of decades: from Markusen's theoretical and methodological 'j'accuse' (2003), to the critical deconstruction of concepts such as 'cluster' (Gordon and McCann 2000; Martin and Sunley 2003), 'regional competition' (Bristow 2005), 'new regionalism' (Lovering 2006; Mac Leod 2001), 'city-region' (Harrison 2007), 'embeddedness' (Hess 2004) and 'local scale' (Mohan and Stokke 2000; Purcell 2006; Purcell and Brown 2005). Perhaps the popularity of the concept of 'industrial atmosphere' in the industrial district literature (Asheim 2003) is the best example of how rhetorically appealing but methodologically magical concepts gained currency in many discourses about local development. It is not by chance that the 'industrial district' is one of the main fields where social capital has been employed extensively as an analytical tool (Farrell and Knight 2003). It is also worth noting that concepts such as social capital and 'industrial atmosphere' have actually made it very difficult to assess the effectiveness of industrial districts in terms of improving the competition of agglomerated firms (Pietrobelli 1998).

Inspired by this critical literature, what this chapter will try to do is to show that this imprecision or ambiguity is not just caused by a lack of analysis and reflection, but is fundamentally rooted in the way in which the notions of social capital and local development refer to the concepts of collective action and collective agency. To illustrate the way in which this magic dispositif works, we need to further engage with the epistemological consequences of the incomplete micro-foundation of social capital and local development theories. In particular, both the notions of social capital and local development are grounded on some form of systemic assumption that 'the whole is more than the sum of its parts'. The question is in what ways do these two approaches justify the 'more'. As previously noted, the concept of social capital has typically been defined with explicit reference to the notion of public good, while commentaries concerning local development often tend to be based on a set of biological, evolutionary and cognitive metaphors.

By first considering social capital, we can start by underlying how Putnam's conceptualization – the one considered to be closest to questions of local development– is far away from Bourdieu's seminal use of the term (see Cannone in this volume). As Bourdieu (1980; 1986) uses it, social capital is part of an individual's assets, together with other forms of capital, mainly economic and cultural. In other words, it is capital in the proper Marxist sense of the term: a possession, either individual or collective, of certain people, who reproduce it by participating in networks and who use it to achieve their own ends. From this perspective, there is nothing magical in his notion of social capital: the very act of networking, in fact, clearly designates here the shift from the individual to the collective. Putnam, instead, argues for social capital as a public good, insisting both that social capital must not be subject to consumption rivalries, and that it must have non-excludable benefits. As in the case of the so-called 'industrial atmosphere', breathing in social capital consequently becomes analogous to utilizing it.

Apart from the fact that Putnam's reference to public goods is superficial and does not consider more recent work examining what public goods are (or could be) (see, for example, Kaul and Mendoza 2003), this account immediately presents some problems with the methodological foundation of the concept itself. Indeed, if we define social capital as a public good we cannot ground its micro-foundation on networking, like in Bourdieu's account. Networks are, by definition, subject to rivalry and exclusion, as is any form of capital. The impossibility of keeping together the network and the public good is clear in much of the literature that shares Putnam's approach:

> Social capital is a feature of the social structure, not of the individual actors within the social structure; it is an ecologic characteristic. In this way, social capital can be distinguished from the concepts of social networks and support, which are attributes of individuals. (Lochner et al. 1999, 260)

The workings of the 'magic' are even more evident if we take into account the way in which Putnam (and his advocates) attempts to empirically ground and test his theoretical hypothesis. By basing his account of social capital on the degree of participation in associations, Putnam's methodology is far from convincing, not only because it might be argued that there could be better proxies (Mohan and Mohan 2002), but mainly because he does not explain how participation in an association – which is a free individual choice – converts into a public good. Furthermore, associations might entail rivalry and exclusion in precisely the same way as in any other social network, since the capital that they produce might not be available to non-members. Putnam's approach is thus not methodologically sound because it attributes the features of a given set of networks to the whole community, regardless of the histories and the geographies of these networks.

Lochner's, Kawachi's and Kennedy's (1999) statement about social capital raises another issue to be considered, since they define social capital concurrently as a public good *and* as an 'ecologic characteristic' [sic]. This definition is extremely important because it sets out a paradox in their usage of social capital. As long as they connote social capital as a public good, they are grounding their reasoning into a well established intellectual tradition of methodological individualism and, more generally, neoclassical economics. When they introduce the notion of 'ecology', and eventually shift to biological metaphors, they abruptly move into the realm of 'hard' sciences in order to naturalize and legitimize their discourse, hence falling into the magic dispositif. It is worth noting that many concepts recurring in local development discourses – from *milieu innovateur* to 'learning region', and from 'localities' to 'innovation system' – are the outcome of the adoption of biological metaphors in economics and also in economic and (mainly) industrial geography. It is well known, in fact, that attempts to establish neoclassical economics as a science have been undertaken by borrowing paradigms from the natural sciences, especially classical physics. A large literature drawing on Georgescu-Roegen's seminal work (1971) critiques this connection between neoclassical economics and classical physics (Lawson 1997; Mirowski 1988). Moreover, such criticisms have significantly been associated with the opinion that biology, and its evolutionary theories in particular, might offer a better framework for the description and comprehension of economic and social processes (Hodgson 1993; 1999).[6]

The work of Trevor Barnes has been especially influential in highlighting the way in which metaphors originating in other sciences have entered geographical discourses. In his *Logics of Dislocation* (1996), Barnes critically documents the diffusion of classical physical and mechanistic metaphors into geography.[7] In a subsequent article, however, Barnes also discusses in positive terms the parallel rise of biological metaphors in economics and geography (Barnes 1998).[8]

The nature and extent of the mechanisms governing this transfer of terms from natural to social science must be considered in order to assess how the magic and loss dispositif works. This transfer can work either as an *analogical* or as a *metaphorical* process, but the difference is significant. By the latter I mean here the adoption of metaphors from one realm to another in order to

6 Despite the contemporary popularity of biological metaphors in economics, we would like to recall how, as long ago as the 1950s, Edith Penrose (1952) sharply criticized some problematic aspects of this intellectual practice.

7 The most (in)famous example of this is the geographical application of the Newtonian spatial gravitation model (see Barnes 1996, ch. 5).

8 In particular, he addresses three potential streams of biological inspiration in economics: (i) Marxist and regulationist theories of reproduction; (ii) evolutionary and institutional economics, with their reliance on the notions of acquired characteristics and inheritance; (iii) and, finally, post-Keynesian economics, with its emphasis on holism and organic structure. The second stream, in particular, has witnessed in the last ten years an explicit attempt to establish an institutional, evolutionary and complex economic geography has been carried out (Amin 1999; Martin 1999 and 2000; Boschma and Frenken 2006; Frenken and Boschma 2007; Martin and Sunley 2007). Moreover, aside from the literature concerned with industrial geography, the use of complexity theory in non-representational geography has established a renewed common ground with biological and ecological discourses (see Thrift 1999; 2005; 2006).

use them as a discursive tool of persuasive rhetoric. By analogical process I mean the transfer of methods – mainly analytical and quantitative – from the natural sciences to social analysis. Apart from a few cases (Pumain et al. 1989), complexity and system theories have been applied to spatial processes but only as sources of metaphoric inspiration rather than as methodological foundations (Plummer and Sheppard 2001). Here, we are driven back once again to the magic micro-foundation of local development: grounding on complexity theories and in particular on cognitive accounts of collective agency, geography is exposed to the temptation of searching a legitimization of its discourses with reference to scientific 'hard' narratives, mainly biological, that cannot be fully translated into social sciences (Byrne 1998).

This is No Longer a Working Number

I do not intend here to praise the rigour of methodological individualism in neoclassical economics or the formal modelling of complexity in 'hard sciences'. Rather I would like to emphasize the risks of an inappropriate use of scientific (falsifiable) methodologies as a shortcut to legitimize unclear magic concepts. The danger here is that through the magical process something gets lost – and this is the second part of the dispositif. When concepts such as social capital and local development are built on slippery or unstable methodologies, they leap blindly from the individual dimension to the collective, generating some 'loss' in our account of the world. If something magic appears, such as a collective subject used to interpret the will, the desires and the needs of the local community, then the need for political negotiation *between* groups *within* the local community in order to direct the development process seems to disappear. Moreover, if collective agency is produced magically through some process of interaction amongst individuals, which is itself magical, then the element of political conflict inherent in consensus-building is lost, and with it, the competing accounts of how public goods ought to be produced and appropriated, and consequently how development paths should be chosen and implemented.[9] In other words, 'methodological magic' is instrumental in building consensus and in marginalizing dissident voices.

---

9 The relationship between the methodological magic and political loss is quite clear in the work of the Italian sociologist Arnaldo Bagnasco and in particular in his account of Putnam's theory of social capital (1999). According to him, the absence of history (that is, the effects of concrete actions of individual and collective actors) from their explanation of social capital leads Putnam's and Fukuyama's accounts both to a deterministic understanding of the relations between social capital and social and economic development (methodological magic) and to downsize the political dimension of social capital, reducing it to a cultural fact. Also Carlo Trigilia, another leading Italian sociologist who was also involved in the conceptualization of industrial district and local development in the 1980s and 1990s, points out the need for both a more rigorous

The translation of social capital and local development theories into mainstream political economy, and in particular into neoliberal agendas, has been readily acknowledged by critical works, mainly with respect to EU policies (Faludi 2004; Hudson 2003; Hadjimichalis and Hudson 2006). Also, on a more theoretical base, the notions of 'rescaling' (Brenner 1999; Swyngedouw et al. 2002) and 'hollowing out' (Jessop 1994) have focused on the fact that the growing emphasis on the regional and local scale is part of a restructuration rather than of a demise of economic and political power relations. My point here is that such attempts to foster a neoliberal vision of development are also perpetrated through the dispositif of magic and loss.

In particular, the employment both of the rhetoric of 'public good', and of biological metaphors is central to the construction of a neoliberal interpretation of the economy and society. In the previous section, I stated that social capital is often conceptualized as a public good but that too little attention is paid to the consequences of this notion. My point is that the reference to the notion of public good is not only based on a magical micro-foundation rooted in methodological individualism, but it also fosters a process of 'economization' of social life, which is often hidden behind the idea of social capital. In particular, these concepts tend to 'lose' their common reference to the history of economic and political thought and they risk becoming ephemeral, rootless words. We might consider, for instance, the apparently obvious breakdown between the 'social' and 'capital'. As I noted earlier in this chapter, for Bourdieu, social capital is first of all a kind of capital – a means of production, reproduced through social networks. The 'social' does not contradict the capital(istic) and the class-related nature of social capital. The public and social dimension is instrumental in the reproduction of capital, but is does not affect the capitalistic organization of society: social capital is not owned socially, it is simply (re) produced socially. Interpreting social capital as a public good, on the other hand, as Putnam does, suggests that it is first of all 'social', and only after that it is a form of 'capital' as well. For Putnam, social capital belongs to the community or to society and not to individuals, so that social capital refers first of all to the 'social' and only subsequently to the 'economical' (1993). From this perspective, Bourdieu's Marxian emphasis on 'capital' is therefore more convincing than Putnam's accent of the 'social', in providing an understanding of how the 'economization' of society works.

The ways in which socio-economic relationships have been reshaped by neoliberalism over the last twenty years are well documented (see for example Harvey 2007). More specifically, neoliberal governmental ideology implies that individuals play an active role in their own control (Dean 1999; Lemke 2001): the emphasis on concepts such as trust, civil society, participation, but also social capital and local development, can be interpreted as a means by which the needs

---

analysis of how social capital is built through networks and a clever account of the political dimension of both social capital and local development (2001).

of global capitalism are met by society (Cooke and Kothari 2001). The real focus here is on a new regulatory dispositif that downscales societal conflicts, which are assumed to be the main obstacles to the success of local economies. Within this process of neoliberal ordering of the world, discourses about collective action and agency, which turn social capital and local development into something methodologically magical, also play a role in undermining the importance of conflict and struggle. When we imagine a trustful local community *à la* Putnam, where individual interests blur into shared, generic stake-holding[10] and collective agency, then we concurrently neglect the fact that real networks of real shareholders activate social capital *à la* Bourdieu in order to attain their own interests. The issue here is not simply justice or inequality, but also the fact that the relationship between the methodological magic and the political loss is often hidden. In this context, the capability of political elites in addressing the local development agenda is safeguarded by holistic accounts of collective agency and, at the same time, the (social) capital of such elites is described as a public good, even when it is clearly not. The use of biological metaphors thus 'naturalizes' existing leaderships, securing consensus and cohesion.

The metaphorical use of terms referring to a kind of 'capital' (be it human, social or, more recently, territorial) leads to surprising consequences. For instance, capital can be discussed without considering thorny issues such as who has it, who instead has no access to it, how it is reproduced and who takes advantage of it. Everyone is assumed to be a 'capitalist' just because he/she lives in a well-endowed place, breathing in a rich industrial atmosphere, with massive doses of social capital pumping through his/her veins. Collective agency – somehow a new embodiment of the Vidalian *genius loci* – works for the common sake, in a similar way to Adam Smith's invisible hand.

Discourses on Regional and Urban Competition: The 'Magic and Loss' Dispositif at Work

Earlier in this chapter, I claimed that both social capital and local development narratives witnessed a sort of ideological u-turn: often evoked as examples of the way in which scholars turned attention onto the significance of the social dimension of the economy, they translated into Trojan horses, which promulgated neoliberal values and attitudes throughout society. I also claimed that the methodological magic (in terms of the way in which collective action and agency are constructed) results in a political loss (since consensus is built and conflict is downplayed), sketching what I called a 'magic and loss' dispositif capable of securing the governmentality of local economies. In particular, I have claimed that the magic and loss dispositif naturalizes consensus,

10 For a critique of the relationship between stakeholder capitalism and local and regional development, see Sunley (1999).

strengthening the control of local and supralocal elites over local processes of development.

In order to consider how this dispositif works, it is useful to look at the question of *regional and urban competition*,[11] that is, the idea that cities and regions compete against each other in order to attract globalized flows of investments, dwellers, events, etc. Regional and urban competition forms a key component in my argument, since it highlights the main features of the magic and loss dispositif for at least two reasons. Firstly, it exemplifies the 'economized' outcome of various discourses on social capital and local development: on the one hand, social capital is understood as being instrumental to competition, while, on the other, the notions of competition and local development become synonymous, such that any trace of the relationship between social capital, local development and resistance to the neoliberal global order is concealed (Conti and Giaccaria 2001). Secondly, it is one of the key concepts through which theoretical and analytical reflections on social capital and local development tend to shift towards policy and normative domains.

Magic

In a recent critical article in the *Journal of Economic Geography*, Gillian Bristow (2005) heavily criticized the concept of regional competition, particularly stressing the weakness of the micro-foundation of the concept. According to her view, the supposed strength of this concept relies on the existence of a mutually reinforcing link between regional features and the behavior of firms: the former is supposed to enhance the competition of firms, whilst their success is claimed to improve regional prosperity. As Bristow points out, this link is often taken for granted, rather than being critically analysed. Despite my sympathy for Bristow's argument, I would like to address the same issue from a different standpoint.

The main question here is not the connection between regions and firms, but rather the role that collective agency plays in regional and urban competition discourses. In other words, can we accept the idea that regions and cities compete in the same way that firms do? Can regions and cities be treated as actors, with their own objectives and strategies? And who actually represents them in competition? The government? The local enterprises? The citizens? More importantly, when we talk about regional and urban competition, we adopt a metaphor that suggests that there is some degree of proximity and resemblance between a region or a city and an enterprise. To what extent can this metaphor be maintained without taking it to its extreme consequences? Are the goals and means of a region similar to those of a firm? Arguably, we

11 These two approaches to territorial competition have been addressed mainly through two special issues of, respectively, *Urban Studies* (Lever and Turok 1999) and *Regional Studies* (Kitson et al. 2004).

have already gone too far in treating cities and regions as firms. For instance, social cohesion is not interpreted as a matter of rights or justice in EU policy-making, but mainly, if not entirely, in order to achieve competitiveness (Sharp 1998; Fainstein 2001; Lawton Smith et al. 2003). In other words, by employing a metaphor of competition, regions and cities are conceived as mere economic actors: social and cultural processes (self-representation, identity, rootedness, cohesion, trust, values etc.) are reduced to the economic dimension of resources, which fuel the competitiveness of cities and regions.

In order to reveal the magic dimension of this process, we can consider how the notion of regional and urban competition is presented by one of the most influential scholars of the Italian approach to local development, Roberto Camagni (2002, 2406):

> If individual firms and individual people undertake collective activities, facilitated by (and creators of) trust and local social capital, if significant cognitive synergies, readily apparent in the local milieu, result from their various interactions and, finally, if these actions and these processes draw additional vitality from cooperation with local public administrations, then it appears justifiable to go beyond methodological individualism – which regards only single firms as operating and competing – arguing the logical validity of a 'collective' concept such as that of territory, and to affirm that territories compete among themselves, using the creation of collective strategies as their instruments.

The presence of the magic dispositif becomes even more apparent if we compare Camagni's statement with the definition proposed by Paul Cheshire and Ian Gordon (1996, 385):

> [Regional and urban competition is] a process through which *groups, acting on behalf of a regional or sub-regional economy*, seek to promote it as a location for economic activity either implicitly or explicitly in competition with other areas. In principle it involves both active local economic development measures of various kinds and a self-conscious strategy to guide policy-development and implementation, with regard both to the future economic role of the area concerned and to its principal competitors. Part of this competitive activity is inevitably addressed to the attraction of investment, sometimes with discrimination between more and less desirable activities to attract. But part can be concerned with enhancing the share (and social profitability) of existing local businesses in the market they serve and generating new businesses and markets. (emphasis added)

In this definition there is a clear analytical concern with the micro-foundation of the concept of regional and urban competition, which is articulated as follows:

i) those who act are groups (of individuals), maintaining a notion of social capital, *à la* Bourdieu, that belongs to some members of the society but not to the society as a whole;

ii) agency works on the basis of a process of representation, 'acting on behalf of something and somebody', which has to be made explicit and cannot be hidden in the black box of collective agency; and
iii) what is 'represented' here is not a region or a city as a whole but rather a 'regional or sub-regional economy'. Hence collective agency is limited to a well defined sphere of the regional social life, i.e. the economy and the production.

My point here is that the question of collective action and agency, which characterizes both the notions of social capital and local development and all discourses on regional and urban competition, underscores the question of the complexity of cities and regions, but it addresses it in a rather ambiguous way. On the one hand, collective agency is supposed to express systemic complexity, postulating that regions and cities are more than the sum of the 'things' that are localized therein, and hence, they possess a set of emergent properties,[12] such as competition (Camagni 2002). On the other hand, the attribution of competition to regions and cities implies the assumption that they 'behave' like economic entities, with some degree of homogeneity in interests, needs, purposes and strategies, an assumption that grossly oversimplifies the complex nature of spatial formations.

Loss

In order to address the (lost) political dimension of regional and urban competition, we can draw on the supposed virtuous circle linking consensus, cohesion and competitiveness. This link is strictly related to questions of 'governance'. The underlying idea here is that regional and urban competition is enhanced by social cohesion and consensus and that, as a consequence, competitiveness can only be implemented through the widespread participation of the greatest possible number of stakeholders: public-private partnerships, real-estate developers, ad hoc agencies for the promotion of foreign direct investments, unions, associations of entrepreneurs, chambers of commerce, influential civic individuals, etc. These broad and diffuse networks of urban and regional actors are supposed to support local administrative authorities in the process of creating new forms of political representation, which might or perhaps should replace traditional national forms of administration of regional economic development. This is of course what the supporters of 'governance' defend (see for example Kooiman 1993). At the same time, we must acknowledge that the notion of governance itself has been greatly criticized for creating collective 'inaction' opposed to action, strengthening existing influential elites within cities and regions (Amin et al. 2000). From a different perspective

12 Contrary to methodological individualism, systemic theory assumes that the aggregate level possesses emergent properties that do not belong to the individuals.

MacKinnon (2000) has developed a Foucauldian interpretation of governance as a form of governmentality, that is as a tool giving central authorities the capability to control the processes taking place at the local scale.

According to these critiques of the concept of governance, the metaphors of 'regional and urban competition' cannot be deemed innocent. In their most extreme expressions these terms imply that whole regions or urban zones are (or should be) organized in such a way as to be competitive. The other implication here is that cities and regions are spatial formations with clear boundaries and a monolithic identity, just like other economic actors such as a group of consumers or a firm. A successful economic and 'territorial' actor is understood as both consistent in its purposes, and organized in its pursuit of those very purposes. This brings us back to the 'juridical personality' paradox that emerges within any holistic approach to the idea of regions and cities, that is, the extreme possibility of attributing to a region or to a city some form of 'collective agency'. More explicitly, talking about 'competition' between regions and cities implies the attribution of a clear economic rationale to a social, cultural and political spatial formation. This is why there is significant degree of consensus regarding the importance of regional and urban competition among both neoliberal policy-makers *and* local development advocates. The former find the question of competition coherent with their 'grand narrative' of economic interest as the engine moving and structuring human relationships, while the latter seem to be more attracted by the emphasis that regional and urban competition puts on community, homogeneity and communality of purposes at the local scale. We can thus fully grasp how, pushing to its extreme consequences one of the main concepts driving the idea/ideology of local development – that is, attributing a collective agency to a region or to a city – can lead to unintended and unexpected conclusions. Indeed, while debates on local development are normally supported by the recognition of the relational, social and cultural dimensions of economic processes, the outcome of regional and urban competition is quite the opposite: conceptualizing and constructing a city or a region as an economic entity.

The point here is that such 'economization of space' is directly related to the naturalization of the working of markets and their spilling over into the whole society. In this process, the adoption of biological metaphors in many approaches to local development is central (Boschma 2004). The concept of competition without doubt comes from a Darwinian account of evolution and therefore it inherits some innate bias towards biological metaphors. Sheppard's account of regional and urban competition offers a useful perspective in considering the relationship between biological metaphors and economic explanations:

> Competition is not only a foundational idea in economic and social theory, but also in biological evolutionary theory. Indeed, over the last century social scientists have frequently appealed to the notions of struggle and selection in Darwinian evolutionary theory to justify the centrality of competition in human societies.

> ... In fact, Darwin borrowed the idea from economics. He was inspired to make struggle and competition central to his evolutionary theory by the economist (as well as population theorist and priest) Thomas Malthus. Darwinian evolutionary theory remains controversial among biologists. ... Alternatives can be conceived. The geographer Kropotkin was among the first to argue that cooperation is pervasive among animals. ... Within economic thinking the discourse of competition is that market-driven (capitalist) competition is generally economically and socially beneficial. This has been articulated through two prevalent metaphors expressing how competition works. The first and dominant one is competition as an invisible hand. ... The second is competition as evolutionary progress. (Sheppard 2000b, 169–70)

Even if we avoid the trap of attributing an economic finality to regions and cities, the holistic biological metaphors which are behind concepts such as regional and urban competition would bring us to another dead end and imply the reification of cities and regions, given once and forever.

The assumption that cities and regions act as an organic whole (as an entity with its own boundaries, able to perform to some degree as a collective agency) has some important consequences. First of all, it implies that regions and cities express a certain degree of homogeneity and cohesion. The local development/social capital debate often tends in fact to assume that they possess an identity, in the twofold sense of having both a unique set of characters which make them different from the external world (the non-local or supra-local) and some degree of temporal continuity, a sort of 'path-dependence' (which, not coincidentally, is one of the recurring terms in local development literature). Once again, theories of local development risk leading to conclusions which are to a certain extent the opposite of their inspiration and starting point. On the one hand, common sense conceptualization of social capital and local development insist on the variety of development paths and on the importance of social and cultural 'biodiversity' (Väliverronen 1998). On the other, when it comes to competition they emphasize internal homogeneity and consensus as 'competitive advantages'.

Another possible unintended consequence of the adoption of regional and urban competition discourses is that conflict is often pushed outside the boundaries of the local community and located in competition against other cities and regions, while 'the local' is praised as the locus of consensus and cohesion. In other words, the 'communitarian' overvaluation of cooperation and shared institutions at the local scale tends to hide the conflictual dimension of the interaction among different groups within the local community.

Conclusions

In this chapter I have argued that the current limits of social capital discourses can be better understood if we contextualize social capital into the broader family of theories and practices that go under the name of local development.

In particular, I have focused on the paradox that, while social capital and local development theories have usually been conceptualized as alternatives to mainstream neoclassical economics, when they are translated into practice they often tend to reinforce neoliberal interpretations of the relationship between the 'social' and the 'economic'. I have also claimed that this paradox depends, at least partially, on the work of a 'magic and loss' dispositif that can be summarized in the following way:

i) both social capital and local development theories are founded on some account of collective action and agency;
ii) in addressing the issue of social action and agency, nevertheless, they miss at least partially the issue of micro-foundation, that is, they rely on a 'magic' explanation of the passage from the individual to the collective;
iii) this 'magic' also produces a 'loss', in particular a loss of political meaning in social capital and local development practices;
iv) the 'magic and loss' dispositif is strictly related to the (mis)use of biological concepts diffused in complexity theory.

Finally I have argued that the rhetoric on regional and urban competition is a useful test-ground for the workings of the 'magic and loss' dispositif in its extreme theoretical and practical implications. The idea that cities and regions compete against each other like firms – or organisms – implies an implicit economization and reification of those very cities and regions. This view assumes that such complex social and spatial formations have a monolithic identity, and that some form of collective action and agency could enable cities and regions, or more generally local communities, to act as a 'juridical person' and hence to compete on a hypothetical 'market of spaces'.

This conflict between theories and practices is a long way off from finding a solution. I strongly believe that the theoretical and methodological foundation of the notions of social capital and local development must therefore be reconsidered in order to avoid this dramatic 'loss of the political' through the narratives of policy-makers. The problem here is that neoliberal organizations like the World Bank or the OECD not only (mis)use the notions of social capital and local development, but that they also develop their own distinct discourses about social capital and local development. Such discourse entails an odd mix of theoretical reflections – often developed in partnership with scholars in an academic setting, and normative practices. More importantly, in such approaches, the concurrence of methodological magic and political loss is straightforward: cohesion, consensus and creativity always fit with competition, productivity and innovation.

The now widespread notion of 'territorial capital' can help in understanding the potential – or maybe the actual(ity) – of this dispositif. The term 'territorial capital' was first proposed not in the academic literature but in the regional policy context.[13] Specifically, it was first used by the OECD in its *Territorial Outlook* (OECD 2001) and then by the DG Regio of the European Commission (EC 2007) in documents on the *Territorial State and Perspective of the EU* (the background document for the Territorial Agenda of the EU). According to the OECD, territorial capital is determined by a series of material and immaterial factors:

> These factors may include the area's geographical location, size, factors of production endowment, climate, traditions, natural resources, quality of life or the agglomeration economies provided by its cities, but may also include its business incubators and industrial districts or other business networks that reduce transaction costs. Other factors may be 'untraded interdependencies' such as understandings, customs and informal rules that enable economic actors to work together under conditions of uncertainty, or the solidarity, mutual assistance and co-opting of ideas that often develop in clusters of small and medium-sized enterprises working in the same sector (social capital). Lastly, according to Marshall, there is an intangible factor, 'something in the air', called the 'environment' and which is the outcome of a combination of institutions, rules, practices, producers, researchers and policy makers that make a certain creativity and innovation possible. (OECD 2001, 15)

It seems in this definition as though all the notions introduced into economic geography over the last 20 years as a result of the cultural and the relational turns have been absorbed indiscriminately. Physical factors (natural resources, accessibility and infrastructures), economic factors (mainly Marshallian external economies), relational features (untraded interdependencies, institutions, practices) and cognitive processes (such as learning, creativity and innovation) all come together grouped in the label of 'territorial capital'. In this way we run the risk of creating a new 'hyper-reified' factor of production – any form of 'capital' is a mean of production – that is implemented without explaining either who *owns* territorial capital and who *takes advantage* of it, or how territorial capital is *accumulated* and *reproduced*, or even who have to be *dispossessed* of their territorial capital in order to allow accumulation.[14] Paraphrasing Cannone's 'Marxian' statement (see Chapter 9 in this book): capital does not melt in the air. Yet what does melt into the air might cause fatal lung disease.

---

13 According to OECD tradition, academic scholars participated, together with the organization's staff, in defining the concept of territorial capital. Among others, Roberto Camagni played a key role in the task (2001). Subsequently, he also tried to translate this notion into a more scholarly recognized framework (2008).

14 See the work of Elyachar about World Bank's Social Fund for Development and the way she applies Harvey's concept of 'accumulation by dispossession' to contemporary neoliberal development discourses (Harvey 2003; Glassman 2006).

Bibliography

Amin, A. (1989), 'Flexible specialisation and small firms in Italy', *Antipode* 21:1, 13–34.

Amin, A. (ed.) (1994), *Post-Fordism. A Reader* (London: Blackwell).

Amin, A. (1999), 'An institutionalist perspective on regional economic development', *International Journal of Urban and Regional Research* 23:2, 365–78.

Amin, A. and Thrift, N. (2000), 'What kind of economic theory for what kind of economic geography', *Antipode* 32:1, 4–9.

Amin, A., Massey, D. and Thrift, N. (2000), *Cities for the Many not the Few* (Bristol: Policy Press).

Arrow, K.J. (1994), 'Methodological individualism and social knowledge', *The American Economic Review* 84:2, 1–9.

Asheim, B.T. (2003), 'Industrial districts: The contribution of Marshall and beyond', in G. Clark et al. (eds).

Bagchi-Sen, S. and Lawton Smith, H. (eds) (2006), *Economic Geography. Past, Present and Future* (London: Routledge).

Bagnasco, A. (1999), *Tracce di comunità* (Bologna: Il Mulino).

Bandura, A. (2000), 'Exercise of human agency through collective efficacy', *Current Directions in Psychological Science* 9:3, 75–8.

Bandura, A. (2006), 'Toward a psychology of human agency', *Perspectives on Psychological Science* 1:2, 164–80.

Barnes, T.J. (1996), *Logics of Dislocation. Models, Metaphors and Meanings of Economic Space* (New York: Guilford Press).

Barnes, T.J. (1998), 'A history of regression: Actors, networks, machine and numbers', *Environment and Planning A* 30:2, 203–23.

Barnes, T.J. (2001), 'Retheorizing economic geography: From the quantitative revolution to the cultural turn', *Annals of the Association of American Geographers* 91:3, 546–65.

Bathelt, H. and Glückler, J. (2003), 'Toward a relational economic geography', *Journal of Economic Geography* 3:2, 117–44.

Boschma, R.A. (2004), 'Competitiveness of regions from an evolutionary perspective', *Regional Studies* 38:9, 1001–14.

Boschma, R.A. and Frenken, K. (2006), 'Why is economic geography not an evolutionary science? Towards an evolutionary economic geography', *Journal of Economic Geography* 6:3, 273–302.

Boulding, K. and Khalil, E. (eds) (1996), *Evolution, Order and Complexity* (London: Routledge).

Bourdieu, P. (1980), 'Le capital social. Notes provisoires', *Actes de la recherche en sciences sociales* 31:1, 2–3.

Bourdieu, P. (1986), 'The forms of capital', in Richardson (ed.).

Boyer, R. and Saillard, Y. (eds) (2001), *Regulation Theory. The State of the Art* (London: Routledge).

Brenner, N. (1999), 'Globalisation as reterritorialisation: The re-scaling of urban governance in the European Union', *Urban Studies* 36:3, 431–51.
Bristow, G. (2005), 'Everyone's a "winner": Problematising the discourse of regional competitiveness', *Journal of Economic Geography* 5:3, 285–304.
Byrne, D. (1998), *Complexity Theory and the Social Sciences. An Introduction* (London: Routledge).
Camagni, R. (2001), 'Policies for spatial development', in OECD.
Camagni, R. (2002), 'On the concept of territorial competitiveness: Sound or misleading?', *Urban Studies* 39:13, 2395–411.
Camagni, R. and Maillat, D. (eds) (2006), *Milieux innovateurs. Théorie et politiques* (Paris: Economica).
Capello, R., Camagni, R., Fratesi, U. and Chizzolini, B. (2008), *Modelling Regional Scenarios for the Enlarged Europe* (Berlin: Springer).
Castelfranchi, C. (2005), 'Cognitive architecture and contents for social structures and interactions', in R. Sun (ed.)
Cheshire, P.C. and Gordon, I. (1996), 'Territorial competition and the predictability of collective (in)action', *International Journal of Urban and Regional Research* 20:3, 383–400.
Clark, G.L., Feldman, M.P. and Gertier, M.S. (eds) (2004), *The Oxford Handbook of Economic Geography* (Oxford: Oxford University Press).
Comaroff, J. and Comaroff, J. (eds) (2000), *Millennial Capitalism and the Culture of Neoliberalism* (Durham, NC: Duke University Press).
Conti, S. and Giaccaria, P. (2001), *Local Development and Competitiveness* (Dordrecht: Kluwer).
Conti, S. and Giaccaria, P. (2006), 'A systemic approach to territorial studies: deconstructing territorial competitiveness', in G. Vertova (ed.).
Cooke, B. and Kothari, U. (2001), *Participation. The New Tyranny?* (London: Zed Books).
Cooke, P. and Morgan, M. (1998), *The Associational Economy. Firms, Regions and Innovation* (Oxford: Oxford University Press).
Crang, P. (1997), 'Introduction: Cultural turns and the (re)constitution of economic geography', in R. Lee and J. Wills (eds).
Dansero, E., Giaccaria, P. and Governa, F. (eds) (2008), *Lo sviluppo locale al Nord e al Sud: un confronto internazionale* (Milan: Franco Angeli).
Dean, M. (1999), *Governmentality. Power and Rule in Modern Society* (London: Sage).
Duranton, G. and Rodriguez-Pose, A. (2005), 'When economists and geographers collide, or the tale of the lions and the butterflies', *Environment and Planning A*, 37:10, 1695–705.
Elyachar, J. (2005), *Markets of Dispossession. NGOS, Economic Development, and the State in Cairo* (Durham, NC: Duke University Press).
European Commission (EC) (2005), *Territorial State and Perspectives of the European Union. Scoping Document and Summary of Political Messages* (Brussels: European Union).

European Commission (EC) (2007), 'The territorial state and perspectives of the European Union: towards a stronger territorial cohesion in the light of the Lisbon and Gothenburg ambitions', <http://www.bmvbs.de/Anlage/original_1005296/The-Territorial-State-and-Perspectives-of-the-European-Union.pdf>, accessed 18 June 2008.
Fainstein, S. (2001), 'Competitiveness, cohesion, and governance: their implications for social justice', *International Journal of Urban and Regional Research* 25:4, 884–8.
Faludi, A. (2004), 'Territorial cohesion: old (French) wine in new bottles?', *Urban Studies* 41:7, 1349–65.
Farrell, H. and Knight, J. (2003), 'Trust, institutions, and institutional change: Industrial districts and the social capital hypothesis', *Politics and Society* 31:4, 537–56.
Frenken, K. and Boschma, R. (2007), 'A theoretical framework for evolutionary economic geography: Industrial dynamics and urban growth as a branching process', *Journal of Economic Geography* 7:5, 635–49.
Georgescu-Roegen, N. (1971), *The Entropy Law and the Economic Process* (Cambridge, MA: Harvard University Press).
Geschiere, P. and Nyamnjoh, F. (2001), 'Capitalism and autochtony: the seesaw of mobility and belonging', in J. Comaroff and J. Comaroff (eds).
Giaccaria, P. and Scarpocchi C. (2008), 'Lo sviluppo locale in Egitto', in E. Dansero et al. (eds).
Giddens, A. (1986), *The Constitution of Society. Outline of the Theory of Structuration* (Berkeley, CA: University of California Press).
Glassman, J. (2006), 'Primitive accumulation, accumulation by dispossession, accumulation by "extra-economic" means', *Progress in Human Geography* 30:5, 608–25.
Gordon, I.R. and McCann, P. (2000), 'Industrial clusters: Complexes, agglomeration and/or social networks?', *Urban Studies* 37:3, 513–32.
Hadjimichalis, C. (2006), 'Non-economic factors in economic geography and in "new regionalism": A sympathetic critique', *International Journal of Urban and Regional Research* 30:3, 690–704.
Hadjimichalis, C. and Hudson, R. (2006), 'Networks, regional development and democratic control', *International Journal of Urban and Regional Research* 30:4, 858–72.
Hadjimichalis, C. and Hudson, R. (2007), 'Rethinking local and regional development: implications for radical political practice in Europe', *European Urban and Regional Studies* 14:2, 99–113.
Hardin, R. (1982), *Collective Action* (Washington DC: RFF Press).
Harrison, J. (2007), 'From competitive regions to competitive city-regions: A new orthodoxy, but some old mistakes', *Journal of Economic Geography* 7:3, 311–32.
Harvey, D. (2003), *The New Imperialism* (Oxford: Oxford University Press).

Harvey, D. (2007), *A Brief History of Neoliberalism* (Oxford: Oxford University Press).
Hess, M. (2004), 'Spatial relationships? Towards a reconceptualization of embeddedness', *Progress in Human Geography* 28:2, 165–86.
Hodgson, G. (1993), *Economics and Evolution. Bringing Life Back into Economics* (Oxford: Polity Press).
Hodgson, G. (1999), *Evolution and Institutions. On Evolutionary Economics and the Evolution of Economics* (Cheltenham: Edward Elgar).
Hudson, R. (2003), 'European integration and new forms of uneven development: But not the end of territorially distinctive capitalisms in Europe', *European Urban and Regional Studies* 10:1, 49–67.
Kaul, I. and Mendoza, R.U. (2003), 'Advancing the concept of global public goods', in I. Kaul et al. (eds).
Kaul, I., Conceição, P., le Goulven, K. and Mendoza, R.U. (eds) (2003), *Providing Global Public Goods. Managing Globalization* (Oxford: Oxford University Press).
Kitson, M., Martin, R. and Tyler, P. (2004), 'Regional competitiveness: An elusive yet key concept?', *Regional Studies* 38:9, 991–9.
Kooiman, J. (1993), *Modern Governance. New Government-Society Interactions* (London: Sage).
Jessop, B. (1994), 'Post-Fordism and the State', in Amin (ed.).
Lawson, T. (1997), *Economics and Reality* (London: Routledge).
Lawton Smith, T., Tracey, P. and Clark, G.L. (2003), 'European policy and the regions: a review and analysis of tensions', *European Planning Studies* 11:7, 859–73.
Lee, R. and Wills, J. (eds) (1997), *Geographies of Economies* (London: Arnold).
Lemke, T. (2001), 'The birth of bio-politics. Michael Foucault's lectures at the College de France on neoliberal governmentality', *Economy and Society* 30:2, 190–207.
Lever, W.F. and Turok, I. (1999), 'Competitive cities: An introduction to the review', *Urban Studies* 36:5–6, 791–3.
Lochner, K., Kawachi, I. and Kennedy, B.P. (1999), 'Social capital: A guide to its measurement', *Health and Place*, 5:4, 259–70.
Lovering, J. (2006), 'The new imperial geography', in S. Bagchi-Sen et al. (eds).
MacKinnon, D. (2000), 'Managerialism, governmentality and the state: A neo-Foucauldian approach to local economic governance', *Political Geography* 19, 293–314.
Mac Leod, G. (2001), 'New Regionalism reconsidered: Globalization and the remaking of political economic space', *International Journal of Urban and Regional Research* 25:4, 804–29.
Magnaghi, A. (ed.) (1990), *Il territorio dell'abitare. Lo sviluppo locale come alternativa strategica* (Milan: FrancoAngeli).

Marchionni, C. (2004), 'Geographical economics versus economic geography. Towards a clarification of the dispute', *Environment and Planning A* 36:10, 1737–53.

Markusen, A. (2003), 'Fuzzy concepts, scanty evidence, policy distance. The case for rigour and policy relevance in critical regional studies', *Regional Studies* 37:6–7, 701–17.

Martin, R.L. (1999), 'The new "geographical turn" in economics. Some critical reflections', *Cambridge Journal of Economics* 23:1, 63–91.

Martin, R.L. (2000), 'Institutional approaches to economic geography', in E. Sheppard and T. Barnes (eds).

Martin, R.L. and Sunley, P. (1996), 'Paul Krugman's geographical economics and its implications for regional development theory: A critical assessment', *Economic Geography* 72:3, 259–92.

Martin, R.L. and Sunley, P. (2003), 'Deconstructing clusters: Chaotic concept or policy panacea?', *Journal of Economic Geography* 3:1, 5–35.

Martin, R.L. and Sunley, P. (2007), 'Complexity thinking and evolutionary economic geography', *Journal of Economic Geography* 7:5, 573–601.

Maskell, P., Eskelinen, H., Hannibalsson, I., Malmberg, A. and Vatne, E. (1998), *Competitiveness, Localised Learning and Regional Development. Specialisation and Prosperity in Small Open Economies* (London: Routledge).

Mirowski, P. (1988), *Against Mechanism. Protecting Economics from Science* (Totowa, NJ: Rowman and Littlefield).

Mohan, G. and Mohan, J. (2002), 'Placing social capital', *Progress in Human Geography* 26:2, 191–210.

Mohan, G. and Stokke, K. (2000), 'Participatory development and empowerment: the dangers of localism', *Third World Quarterly* 21:2, 247–68.

OECD (2001), *OECD Territorial Outlook* (Paris: OECD).

OECD (2002), *Local Economic and Employement Development* (Paris: OECD).

OECD (2007), *Globalisation and Regional Economies. Can OECD Regions Compete in Global Industries?* (Paris: OECD).

Olson, M. (1965), *The Logic of Collective Action: Public Goods and the Theory of Groups* (Cambridge, MA: Harvard University Press).

Ostrom, E. (1990), *Governing the Commons. The Evolution of Institutions for Collective Action* (Cambridge: Cambridge University Press).

Ostrom, E. (2000), 'Collective action and the evolution of social norms', *The Journal of Economic Perspectives* 14:3, 137–58.

Ostrom, E. and Ahn, T.K. (eds) (2003), *Foundations of Social Capital* (Cheltenham: Edward Elgar).

Overman, H. (2004), 'Can we learn anything from economic geography proper?', *Journal of Economic Geography* 4:5, 501–16.

Penrose, E.T. (1952), 'Biological analogies in the theory of the firm', *The American Economic Review* 42:5, 804–19.

Pietrobelli, C. (1998), 'The socio-economic foundations of competitiveness: An econometric analysis of Italian industrial districts', *Industry and Innovation* 5:2, 139–55.

Pike, A., Rodriguez-Pose, A. and Tomeney, J. (2007), *Local and Regional Development* (London: Taylor and Francis).

Plummer, P. and Sheppard, E. (2001), 'Must emancipatory geography be qualitative? Reply to Amin and Thrift', *Antipode* 33:2, 194–9.

Porter, M. (1990), *The Competitive Advantage of Nations* (New York: Barnes and Noble).

Pumain, D., Sanders, L. and Saint-Julien, T. (1989), *Les villes et auto-organisation* (Paris: Economica).

Purcell, M. (2006), 'Urban democracy and local trap', *Urban Studies* 43:11, 1921–41.

Purcell, M. and Brown, J.C. (2005), 'Against the local trap: scale and the study of environment and development', *Progress in Development Study* 5:4, 279–97.

Putnam, R. (1993), *Making Democracy Work. Civic Traditions in Modern Italy* (Princeton, NJ: Princeton University Press).

Putnam, R. (2000), *Bowling Alone. The Collapse and Revival of American Community* (New York: Simon and Schuster).

Richardson, J. (ed.) (1986), *The Handbook of Theory. Research for the Sociology of Education* (New York: Greenwood Press).

Scott, A.J. (1998), *Regions and the World Economy. The Coming Shape of Global Production, Competition and Political Order* (Oxford: Oxford University Press).

Scott, A.J. and Storper M. (2003), 'Regions, globalization, development', *Regional Studies* 37:6–7, 579–93.

Sharp, M. (1998), 'Competitiveness and cohesion: Are the two compatible?', *Research Policy* 27:6, 569–88.

Sheppard, E. (2000a), 'Geography or economics? Contrasting theories of location, spatial pricing, trade and growth', in G.L. Clark et al. (eds).

Sheppard, E. (2000b), 'Competition in space and between places', in E. Sheppard and T. Barnes (eds).

Sheppard, E. (2001), 'How economists think. About geography, for example: Reflections on the spatial economy', *Journal of Economic Geography* 1:1, 131–6.

Sheppard, E. and Barnes, T. (eds) (2000), *A Companion to Economic Geography* (Oxford: Blackwell).

Storper, M. (1997), *The Regional World. Territorial Development in a Global Economy* (New York and London: Guilford Press).

Sun, R. (ed.) (2005), *Cognition and Multi-Agent Interaction. From Cognitive Modeling to Social Simulation* (Cambridge: Cambridge University Press).

Sunley, P. (1999), 'Space for stakeholding? Stakeholder capitalism and economic geography', *Environment and Planning A* 31, 2189–205.

Swanson, G.E. (1992), 'Doing things together: Some basic forms of agency and structure in collective action and some explanations', *Social Psychology Quarterly* 55:2, 94–117.

Swyngedouw, E., Moulaert, F. and Rodriguez-Pose, A. (2002), 'New geographies of power, exclusion and injustice. Neoliberal urbanization in Europe: Large-scale urban development projects and the new urban policy', *Antipode* 34:3, 542–77.

Thrift, N. (1999), 'The place of complexity', *Theory, Culture and Society* 16:3, 31–69.

Thrift, N. (2005), *Knowing Capitalism* (London: Sage).

Thrift, N. (2006), 'From born to made: Technology, biology and space', *Transactions of the Institute of British Geographers* 30: 463–76.

Trigilia, C. (2001), 'Social capital and local development', *European Journal of Social Theory* 4:4, 427–42.

Udehn, L. (2001), *Methodological Individualism. Background, History and Meaning* (London: Routledge).

Väliverronen, E. (1998), 'Biodiversity and the power of metaphor in environmental discourse', *Science Studies* 11:1, 19–34.

Vertova, G. (ed.) (2001), *The Changing Economic Geography of Globalization* (London: Routledge).

Yeung, H.W. (2005), 'Rethinking relational economic geography', *Transactions of the Institute of British Geographers* 30:1, 37–51.

World Bank (2004), Local Development Discussion Paper, prepared for the International Conference on Local Development (Washington, 16–18 June 2004) (Washington DC: World Bank).

World Bank (2006), *Local Economic Development. A Primer Developing and Implementing Local Economic Development Strategies and Action Plans* (Washington DC: World Bank).

PART 2
Grounding Networks

Chapter 4
Rescaling Trieste: (Not So) Invisible Networks, (Dis)trust and the Imaginary Landscapes of the Expo 2008

Giuseppe Porcaro and Claudio Minca

Introduction

Trieste's (failed) bid for the Expo 2008 is often described as a defining moment in the recent history of this Adriatic city. It represents an 'urban episode' that brought to light, in many ways, all the limits of a putative Triestine social capital,[1] but that also made evident the fact that international competition among cities cannot rely solely on images and presumed 'urban traditions' exhumed for the occasion, or on the mere logics of 'grant coalitions' – that is, *ad hoc* networks that come together for a specific bid, but lacking a longer term vision and a solid degree of 'trust' that an urban milieu characterized by the strong presence of social capital is normally capable of mobilizing.

The argument that we advance in this chapter is that Trieste's failed bid for Expo 2008 was due not only to a set of unfavourable 'geopolitical' conditions that privileged a different choice (despite the apparent consensus that the Trieste candidacy gathered during the process), but also related to the city's lack of a series of reliable material and immaterial networks, potentially capable of mobilizing fiduciary relationships and a longstanding tradition of local 'working togetherness' – elements that are often identified as the outcome of the enactment of a strong, well established collective social capital. This perception, we believe, has diminished the effects of the considerable efforts of many of the city's institutions and Expo organizers, and of a relatively strong consensus among the citizenry.

We take here social capital as the capacity of local networks of trust to be mobilized for a specific project (see Cannone's contributions in this volume). The spatial dimension of social capital and trust – which is the main focus of this collection – is analysed here in the light of the failed bid. This is an episode that

1 See the discussion on social capital in the Introduction and in Chapters 2 and 3 of this volume.

proved to be particularly useful in order to 'test' the workings and the 'added value' provided by precisely those intangible networks of trust and territorial competence that make possible the creation and enactment of urban social capital. It is on occasions such as this, in fact, that the city's elite and its driving ideology are forced to engage with diverse operational and cognitive scales (see Giaccaria in this volume) that are often distant from the conventional structures of self-representation produced and supported by local urban actors.

By adopting a mainly geographical perspective on these issues, we will thus explore the cultures, the scales and the discourses that have characterized the construction of a specific image of Trieste as a hypothetically ideal host city for an international event of such breadth. The city's bid for the Expo between 2003 and 2004 provides, in fact, a useful example of how new networks and local discourses can be forged around a specific idea of a city, its future development and its external projection, and how social capital is (or is not) implicated in this.

Figure 4.1 Trieste's waterfront (photo by Claudio Minca)

Like many other hallmark events, Expos are often considered concrete 'policy vehicles' capable of producing new urban spatialities, improving a city's image and enhancing and consolidating its role within international urban networks. Some southern European cities have managed to successfully translate such events into much wider initiatives of urban development – it is enough to recall, for example, the success of the Olympics in Barcelona, the Expo in Seville and, at least in part, the celebrations of the 500 years of Columbus' journey in Genoa.

This trend is often presented as a sort of postmodern rediscovery of the glorious days of the International Exhibitions of the past, whose legacy still carries symbolic images of the Crystal Palace and the Eiffel Tower (see Billinge 1993; Chalkey and Essex 1999; De Spuches 2002). In this chapter, we will conceive the Expo as a potential collective agent of urban change, operating at multiple scales and through diverse discursive formations (i.e. narratives and mythologies of places and spaces). The Trieste bid for Expo 2008 was indeed interpreted by some representatives of the urban elite and also by some institutions as a tool for the 'rescaling' of the urban landscape and of the city's geographical imaginations, through the selective mobilization of metaphors, people and networks.

In this regard, an examination of Trieste's candidature is particularly revealing. Trieste adopted a well known recipe: the transformation of the Porto Vecchio (the Old Port) into a regenerated exhibition area to be 'given back to the city' once the Expo had finished. Furthermore, the project tried to link the event with a set of new investments aimed at the internationalization of the city and its image. There is nothing specifically new in this strategy, at least if compared to other similar experiences, for example, in Barcelona, Athens or Sydney. However, the negative outcome of Trieste's bid and the specific relationship between local networks and geographical metaphors that the whole process highlighted allows us to bring a different perspective to the discussion on the place-specific nature of social capital, especially when confronted with other well-known and successful 'best practice' cases.

Figure 4.2 The Porto Vecchio (photo by Claudio Minca)

For this reason, we will focus our attention on both the 'materialities' of the Trieste bid – such as the plans for urban renovation – and the 'invisible cultures' of the Expo, constituted by the intangible networks and the geographical and historical metaphors about 'Trieste' and the Triestini mobilized for the occasion. Within this perspective, we will critically assess the political and cultural milieu within which the idea of the Expo originated, taking into account the renewed interest in urban regeneration at the institutional level since the end of the 1980s, often 'imagined' through various projects for the revitalization of the Porto Vecchio. As for the 'invisible cultures' that were evoked in support of the bid, we will instead take into consideration narratives concerning Trieste's supposed 'urban centrality', together with some key metaphors adopted for the international promotion of the city and its urban transformation and the 'theming' of the event in the bidding documents and during the official presentations to the Bureau International des Expositions in Paris, between June 2003 and December 2004. Finally, we will discuss the 'geographical interplay' of the stakeholders involved in the bidding process and their role in the 'politics of scale' identified in this particular context, in order to try to link the failure of the bid to the 'lack' of the sort of urban social capital necessary to the success of such an event.

'A Midsummer Night's Dream': The Expo and the Rescaling of Local Governance in Trieste

In Trieste, city marketing policies and the renovation and gentrification of run-down and de-industrialized areas of the city mirror, in many ways, the initiatives of other European cities (see Colombino and Minca 2005). Nonetheless, Trieste's urban strategies have to contend with the challenge posed by the ghosts of a unique and prestigious past, that of a city entirely re-invented by the Habsburg Empire and transformed into a booming, cosmopolitan Adriatic hub for Vienna's oriental ambitions (see Minca in this volume and Minca forthcoming, but also Ara and Magris 1982). An important symbol of Trieste's past as the erstwhile port of the Empire is the Porto Vecchio, with its strategic location at the fringe of the city-centre and an architecture that poignantly evokes the glorious past of the Adriatic city.[2]

We therefore consider the Porto Vecchio as a crucial non-human actor in the production of Trieste's 'social capital', also today. Much of the material and symbolic capital of the city has, in fact, been built around its self-representation as a port-city, and the port's infrastructures have played a key role in this representation (and function). The docks, indeed, acquired new meaning when,

2 The creation of a free port in Trieste was part of the Austro-Hungarian Empire's attempt to play a leading role in maritime commerce in the Adriatic, until then the monopoly of Venice and the Ottoman Empire (see Rumpler 2003).

after its golden days, the Porto Vecchio underwent a period of decay. Since the 1970s, as a consequence of the technical and logistical changes in the nature of maritime traffic, the existing infrastructures have become redundant – confirmed by the very limited activity still taking place there (see Minca 1995). Most of the port-related activities have been relocated to the New Port, on the southern outskirts of the city, and kept spatially separate from the life of the city centre. During the last few decades, however, many projects for the revitalization of the urban waterfront (including the Old Port) have been put forward in order to find a new role for this beautiful but semi-abandoned part of the city centre. However, despite the fact that the debate over the future of Porto Vecchio has marked the political debate in Trieste for generations now, the old docks remain substantially unused and they represent, in the perception of many, an incredibly valuable symbolic and functional heritage for the city. It is in this context that the bid for Expo 2008 emerged, with plans for the redevelopment of the Porto Vecchio as not only the spatial strategy for its success, but also a legacy to be bequeathed to the city.

The first project for reviving the area was presented in 1971, as part of a broader set of guidelines for the redevelopment of Trieste by Japanese architect Kenzō Tange (see Tange and Urtec 1971).[3] In 1974, another project was presented by Italian architect Guido Canella. His plan was to maintain some of the port's maritime activities, while also imagining a set of innovative land uses, such as an urban park, spaces for fairs, exhibitions and scientific activities and a large parking lot. A number of other projects were proposed at the beginning of the 1990s. Among the most relevant was a project entitled Bonifica, supported by the Friuli-Venezia Giulia Region (and prepared by the Iri-Italstat Institute), and the Polis project, jointly promoted by the Port Authority, Fiat and the Generali insurance company (a key economic actor in Trieste). By the mid-1990s, the prospect of redeveloping this area attracted growing interest on the part of both private investors and public institutions: in a city where space was at a premium, the real possibility of regenerating an entirely new part of the urban centre was increasingly appealing. The debate around the future of the Porto Vecchio was also fostered by the new municipal government elected in 1993 – a government with new-found legitimacy thanks to a new electoral law that allowed, for the first time, for the direct election of a mayor.

In this renovated political context a new proposal, Trieste Futura, was presented by the local Chamber of Commerce and other relevant representatives of the local economy and civil society. This new proposal took shape between 1997 and 1999, with a strong involvement of the citizens in a series of public debates, marking an important break with previous strategies which had been largely based on external intervention and financial support (see Trieste

3 Kenzō Tange's team imagined a comprehensive redevelopment of the coastline, with motorways and skyscrapers along the waterfront between Sistiana Bay and the urban centre.

Futura 1998a).[4] The project entailed the complete transfer of the remaining port activities to the new docks on the eastern fringe of the city and a complete renovation of the landscape and the functions of the Porto Vecchio. Drafted by Manuel de Solà Morales, who played a role in the regeneration of the Port Vell in Barcelona before the Olympic Games in 1992, the new waterfront imagined by Trieste Futura reflected in many ways the models previously realized in Baltimore and Barcelona: a highly aestheticized and gentrified environment including marinas, hotels, cafés and entertainment facilities, accompanied by research facilities and office space for both private and public companies and institutions. The project also aimed to improve traffic flow along the waterfront, envisaging the realization of a new access route to the city centre within the area of the harbour, in line with the expected increase in traffic in that part of Trieste.[5] However, the project lacked the necessary legal backing in order to move to the implementation phase and, like the ones that preceded it, was eventually abandoned. Nonetheless, it succeeded in igniting a lively public debate on the future of the Porto Vecchio and in highlighting the enormous potential of the area, as well as the role it could play in the future of the city.

A few years later, the Port Authority supported a new, though in many ways similar, initiative by proposing to change the Piano Regolatore Portuale (the Port Master Plan), creating a public-private partnership, the Porto Vecchio Srl., aimed at implementing a new regeneration project (see Autorità Portuale di Trieste 2001). This project did not foresee any major change in the maritime activities of the old docks but aimed, instead, to combine them with a more general urban redevelopment of the area. Furthermore, among its objectives was the creation of a new port-related company, a sort of incubator for the coordination of logistic activities for shipping, benefiting from the privileged regime of the free port which still applied to this area (see Boeri 2001).

In between those two proposals, and somehow connected to them, the idea of Trieste's candidacy for Expo 2008 started to take shape, also building on the momentum created by the debates mentioned above. The idea of an International Exhibition in Trieste had already begun circulating in 1998, promoted by some local academic and professional circles (see *Il Piccolo* 1998). Soon, the Expo became something more than just a 'midsummer night's dream' (as it was initially defined by its promoters), with the constitution of an *ad hoc* agency, the TriestExpo, and a set of collaborative initiatives with the newly

4 The Trieste Futura association was established in 1997 by Federico Pacorini, then President of Trieste's Chamber of Commerce, with the participation of many other associations, including professional and entrepreneurial groups, and in close cooperation with the municipality.

5 The project foresaw the construction of an underground parking lot for 450 cars. The plan also called for a reflection on the logistical platform for the development of the commercial port (see Trieste Futura 1998b).

Figure 4.3 Another view of the Porto Vecchio (photo by Claudio Minca)

created Faculty of Architecture at the University of Trieste.[6] According to this group of professionals, Trieste needed an integrated approach and strong guidelines capable of attracting the attention of key decision makers and of potential investors, also in order to bring together the resources available for the realization of new strategic urban development. For this reason, the proposal foresaw the involvement not only of many experts in planning but also in communication and urban governance. In line with this approach, the feasibility study, officially presented in 2000, illustrated a project not limited to the development of the Porto Vecchio, but spanning the entire Trieste Province and Friuli Venezia-Giulia Region, with direct links also with the bordering areas of the Veneto Region and Slovenia.

The idea of regenerating the Porto Vecchio as part of the Expo initiative worked in synergy with the projects for a Trieste Futura. This new 'integrated' approach also tried to engage with the needs of other urban actors. One example was the idea of a *No-car Expo* which, besides the attempt to solve the logistical problems specifically related to the event, was linked to the need to solve the question of urban mobility through investments in public transportation and the construction of a tram line along the waterfront. Another idea was that of a *Floating Expo*. The exhibition areas were to be located on floating structures

6 Exploratory work in the summer of 1998, conducted by Professor Giuseppe Borruso and his team, included visits to exhibition sites in Seville and Lisbon and to the Headquarters of the *Bureau International des Expositions* in Paris (see Borruso 1999).

along the docks, thus allowing for the 'proper' regeneration of the old port buildings while avoiding the risk of damaging them through the presence of massive numbers of visitors. Finally, the theme of mobility, the key *leit motiv* of this initiative – sea-routes, sky-routes, land-routes – recalled the potential strategic role of Trieste as an international transportation hub (see Carapetian and Caputo 2000).

Figure 4.4 Empty warehouses in the Porto Vecchio (photo by Claudio Minca)

The projects for TriestExpo were frozen for a period in 2002, following a change in the municipal government.[7] Nevertheless, the new administration soon decided to take over the idea. Trieste was thus officially put forward as a candidate for the Thematic Exhibition of 2008, although the original proposal was partially abandoned. A new public-private agency was created to manage the bid – the Trieste Expo Challenge 2008 SCpA – with the active presence and support of the Municipality, the Province and the local Chamber of Commerce. The board of the company was, at a later stage, enlarged to include representatives of the Regional Government and the Port Authority. At a diplomatic level, the bidding process was managed by a dedicated committee

7 With the transfer of the bid's management to a new public-private initiative, the TriestExpo together with the Trieste Futura created in 2003 a new project: Triesteidea, aiming to foster a new debate on the future of the city (see Borruso 2004).

created by the Ministry of Foreign Affairs, in close cooperation with Trieste Expo Challenge.[8] The candidature was presented to the Bureau International des Expositions (BIE), on 12 December 2003.[9]

The new agency quickly put together a novel master plan for the Expo area, to be annexed to the bidding documents. As in the previous proposal, the Porto Vecchio was considered the best available site for the event, though this time the proposal was to use the old buildings themselves as exhibition spaces. New structures such as a Conference Centre and an Intercultural Tower were to be created (see Trieste Expo Challenge 2004a). In addition, the project included a post-Expo renovation plan – not very different from the previous proposals – placing strong emphasis on the idea of 'bringing the Porto Vecchio back to the city', an idea that at this point had been accepted also by political actors traditionally more concerned with the survival of some key maritime activities in that area. Nevertheless, the plan was largely a patchwork of previous projects and seen as a limited intervention in the port area. It definitely lacked a strong strategic vision and an integrated approach to the rest of the city's urban development. This weakness emerged quite clearly in the bidding documents, with some of the project sections appearing to be more of a catalogue of all previous planning documents approved in Trieste in the past years, rather than representing a sea-change in the vision for the city and its future.[10]

In the end, Trieste competed for the bid with the theme *The Mobility of Knowledge*, running against the candidatures of Thessaloniki (focused on sustainable development, with the theme *Terra Mater*), and Zaragoza (with *Water* as its guiding theme). There were several key moments in the history of the bidding process: the presentation of the candidatures to the BIE (June 2003 –December 2004); the official presentation of the bid documents (February 2004); the BIE mission's visit to Trieste (March 2004); and throughout, an intense diplomatic activity aimed at galvanizing the necessary international support for the initiative (see Pichierri and Pugliese 2004). The competition was officially closed on 17 December 2004 in Paris, when Zaragoza, with 57 votes out of 91 delegates of the BIE Assembly, was awarded Expo 2008 (see Taino 2004).

With the failure of the bid, the entire redevelopment plan was abandoned. Nonetheless, it is useful to reflect on the nature and the impact of this failed

8 Giuseppe Porcaro's interviews with Fabio Assanti, President of Trieste Expo Challenge, and Massimo Greco, vice-President of the Trieste Province, 19 July 2004.

9 The official candidature was accepted by the BIE in response to the request made on the 23 May 2003 by then Italian Minister of Foreign Affairs, Franco Frattini. The first official presentation to the BIE Assembly was made in June 2003.

10 The bidding documents mention different ongoing and planned urban projects without any specific link to a strategic and coherent vision for urban development and the impact of the Expo. Among the projects cited is European transportation Corridor V, an underground tunnel, and other plans for urban mobility and urban renovation (see Trieste Expo Challenge 2004a).

initiative. In many ways the bidding process represented not only an attempt to put Trieste 'back on the map' but, what is more, at the very 'centre' of the map, as so many of the representations of the Porto Vecchio's presumably global and European 'centralities' highlighted. In so doing, the actors involved in the bidding process mobilized a distinct set of geographical metaphors: those of cosmopolitanism and multiculturalism, 'Mitteleuropa' and 'the Mediterranean'. In what follows, we explore the ways in which such specific representations of the city and the Porto Vecchio were produced, as well as some of the reasons for their failure to link a supposedly consensual local ideas about the city with the ways in which Trieste and its bid were perceived externally – and why the lack of well-defined social capital was crucial in that respect.

Geographical Imaginations

A red 'X' that embraces the globe; two red, white and green trails (the colours of the Italian flag) left by two planes travelling around the world; two meridians crossing above Trieste: this was the symbolic intended meaning of the logo for Trieste's Expo 2008 bid. The informational brochure distributed in March 2004 to the local population highlighted a very specific geographical imagination: the 'X' logo representing the city's determination to rediscover its central role as 'a crossroads for peoples and cultures, goods and ideas'.

The imagination of a new international centrality for Trieste was indeed at the heart of the Expo idea. Already in the first stages of the project elaborated by the TriestExpo team, the Expo was envisioned primarily as a strategic tool for bringing the city back on the international stage. The regeneration of the waterfront and of the Porto Vecchio were presented as key functional elements in a broader project that would reclaim a strong central role in the European context for a city that, for quite some time, had found itself not only at the peripheries of the national economy but also at the peripheries of Europe, its fate marked by its location on the Cold War border. This focus on Trieste's 'new centrality' was a crucial aspect in the promotional strategies of the institutional actors. In this sense, the documents produced by the Trieste Expo Challenge during the bidding process provide a useful starting point for our analysis of the ways in which Trieste's image was built through the mobilization of a distinct set of geographical metaphors. The documents examined here are of three types: (i) promotional materials – brochures and informational booklets distributed at the local level; (ii) the bidding dossier – presented in February 2004, both in English and French, for an international readership; and (iii) the videos produced for the four official presentations at the headquarters of the Bureau Internationale des Expositions in Paris. Analysis of these documents was supplemented by interviews with the coordinators of the Trieste Expo Challenge and, in particular, with the Challenge's press and communication officers; a number of local journalists were also interviewed, in particular those

Figure 4.5 Porto Vecchio and the corporate image of Trieste Expo 2008 (© Trieste Expo Challenge 2008)

working for Trieste's daily newspaper *Il Piccolo*, a key political voice in the city, and for the local TV broadcaster Tele 4.[11]

As suggested previously, the strategy put into place by the Trieste Expo Challenge was not that different, either in form or content, from city marketing campaigns launched by numerous other cities around the world. Just as in countless other cases, the Trieste bid relied on a set of images of a beautiful waterfront awaiting regeneration, an emphasis on the potential international relevance of the city and the 'uniqueness' of its local context. At least three key axes can be identified in the representation of the city, emphasizing its supposed international centrality on the basis of certain historical, geographical, architectural and political characteristics (these are summarized in Table 4.1).

The first trope/axis here is the supposedly 'multicultural' character of this border city – quite a common argument advanced by many European cities today (see Colombino forthcoming). In the Trieste case, multiculturalism is presented as the outcome of the city's 'vocation' as a *cultural crossroads*, as claimed by an informational brochure distributed in 2004 as well as the bidding dossier. This trope draws on the city's history as a 'cosmopolitan' port of the

11 The interviews took place on 20 and 21 July 2004, in the headquarter of *Il Piccolo* (with Paola Bolis) and of *Iniziative Culturali* (with Roberto Morelli).

Table 4.1 Representations of Trieste's 'centrality' in the bidding process

Axis	Brochures	Bidding document	Presentations to the BIE
Cultural crossroads	'Trieste has been, for centuries, a harmonious meeting place for diverse ethnicities' (2002) 'Crossroads of cultures … open to dialogue and communication' (2004a) 'The entire city can be considered a place able to attract knowledge' (2004a)	'A cross-roads of different cultures' (p. 7) 'Cosmopolitan city' (p. 22)	'One of the most significant melting pots of Europe' (June 2003) 'Cosmopolitan centre of Europe' (June 2003)
Knowledge 'pole'	'Trieste, capital of science' (2004b) 'International Centre for Science and Technology' 'Trieste is one of the key hubs between the North and the South, the East and the West, between developments in the Mediterranean and the future of Europe' (2002)	'Trieste, symbol of the mobility of knowledge' (p. 41) 'Trieste as a "knowledge port"' (p. 41)	'The "city of science" is that part of Trieste projected into the future' (June 2003) 'Plateforme pour l'exchange de la connaissance' (December 2003)
Economic hub	'Trieste is a model for development' (2004b) 'Thanks to the new political configuration of the Continent, it will again be the centre of Europe' (2004c)	'A new role for the city and its territory in a European and Mediterranean context which is constantly expanding with the enlargement of Europe to the East' (p. 21) 'Trieste could play a pivotal role connecting the economies of the Danube and Mediterranean regions' (p. 23)	'If you want to draw a circle around the whole of Europe and the area of the Mediterranean you would have to point the needle of the compass at Trieste' (June 2003) 'Trieste est au centre de l'Europe, le site est au centre de Trieste' (December 2003) 'If the world is a global village in 2008 Trieste will be again its market square' (June 2004)

Austrian Hungarian Empire and a unique model of urban cohabitation and coexistence (see Minca in this volume); this representation was particularly visible in the presentations to the BIE, often conflating in quite problematic ways ideas of 'multiculturalism' and 'cosmopolitanism' (for a critique, see Beck 2006). The second trope/axis presents Trieste as a 'knowledge pole', in recognition of the city's formidable scientific community and institutions. Such a trope was directly linked to the proposed theme of the Expo, the mobility of knowledge, a theme of great strategic relevance for the city's future, in particular given Trieste's aim to bring together a 'port of trade' with a 'port of ideas'. The third trope identifies Trieste as a putative 'economic hub', thanks to its 'geographic position'. In both the brochures and the dossier for the bidding it is clearly stated that the city proposes itself to the world as a potential, almost natural, focal economic point between the South and the North, the East and the West, the Mediterranean and continental Europe, especially following EU enlargement.

This supposedly 'natural' geographical centrality was literally put on stage through a theatrical performance during the first presentation of the bid in June 2003 in Paris. On that occasion, a short film was projected, depicting some of the most famous tourist sites in Italy by means of a virtual journey from the South to the North, ending in Trieste. The fictional journey was actually the personal itinerary of a woman travelling to visit her partner in Trieste – a storyline straight out of a TV soap opera, rather than urban promotional literature. At the end of the film's projection, one of the actors appeared on stage, declaiming theatrically that should the delegation 'wish to draw a circle around all of Europe and the Mediterranean, they would have to point the needle of the compass at Trieste' – a claim that had already been made by at least ten other cities in the previous decade!

These examples provide a good illustration of some aspects of the communicative strategy and objectives adopted by the Agency promoting the bidding process. The images and projects were, in fact, selected by Trieste Expo Challenge with no apparent concern for potential conflicts and contradictions between the various representations, but also between the various social/territorial actors and networks presumably represented. For this reason, it is important to critically engage with some of the geographical tropes presented by the marketing strategy for the bid, and analyse these with reference to the wider contexts which they claimed to evoke.

Trieste's 'Global' Centrality

Trieste's 'international centrality', as presented by the Trieste Expo Challenge, was based, first of all, on the city's 'geographical position', envisioned as a sort of 'natural' interface between global and European networks. In concrete terms, this supposed 'natural' centrality was built upon the historical role of the port which, until the fall of the Habsburg Empire, was an important international marketplace and one of the main hubs for trade between Central Europe and the rest of the world. As highlighted by Minca elsewhere in this volume, Trieste's privileged position radically changed after World War I with its incorporation into the Italian kingdom and the radical geopolitical reconfiguration of the former Empire. This new peripheral condition became even more consolidated after World War II, when the city found itself located just a few kilometres away from the Iron Curtain. After the end of the Cold War and the ensuing enlargement of both the European Union and, subsequently, the Schengen space, this 'constrained' peripheral condition has, in many ways, disappeared and Trieste can aim at a different role in the 'new Europe'. What is more, the globalization of the shipping trade offers further opportunities: the city could potentially benefit from its competitive advantage over other European ports in controlling Asiatic routes, whilst the Suez Canal allows goods to arrive more quickly to the Northern and Eastern European markets if they pass through Trieste (see Caroli et al. 2003). These, among others, were the arguments presented during the bidding process.

However, the situation of Trieste's port was miles away from its representation in the bid documents. After a period of decline, the port was still suffering from both logistical and structural weaknesses, seriously hampering its competitiveness in global markets. Such difficulties were, indeed, the focus of on-going public debates in the city. Nevertheless, the bidding documents reflected none of these challenges – nor proposed alternative solutions. The relatively poor performance of the harbour was not only due to scarce local competitiveness, but also linked to the changing global strategies of maritime companies and to the lack of cooperation between Adriatic ports. Increasingly, the 'ranking' of ports is linked directly to the policies of large shipping companies that manage docks in a number of locations.[12] In June 1999, a new port system for the Northern Adriatic (NAPAN) was established, as part of the Central European Initiative (CEI).[13] NAPAN's main aim was to further develop linkages promoted by the European Union, in particular by fostering transborder cooperation between the ports of Koper (Slovenia), Rjieka (Croatia) and Trieste. The concrete objective of the initiative was to improve the maritime transport infrastructure of the region and to accelerate the economic and social development of the coastal areas through a variety of

12 For instance, the arrival of the MSC cruise ship *Viviana* in July 2004 blocked all the ordinary activities of one of the docks, the *Molo VII*. The incident sparked an open debate between local politicians and entrepreneurs over the future of the Port. Previously, the Chinese shipping giant COSCO (China Ocean Shipping Company) had also abandoned Trieste, choosing the Port of Naples for its activities instead (see Contessa 2004a/b).

13 The Central European Initiative (CEI) was established in 1989 as an intergovernmental forum for economic, political and cultural cooperation between Eastern and Central European states and the European Union (see <http://www.ceinet.org>). The secretariat of the CEI is in Trieste.

projects financed by Stability Pact Funds and other EU funding mechanisms (see Pritchard 2004). Despite the evident advantages of increased cooperation,[14] the plans for integrating the ports of Koper and Trieste came to a halt because of opposition from several influential local politicians in Trieste, who argued that Luka Koper (the company in charge of the port activities in the Slovenian city) would work to deflect traffic to the Slovenian harbour at the expense of Trieste. The failure of this cooperation initiative also created a deadlock in the integration strategies between other ports of Northern Adriatic.

This abortive attempt at integration should also be seen, however, as the fruit of a broader set of problems plaguing Trieste's role in continental transportation networks, including its lack of a viable logistical platform, an efficient intermodal system and suitable connections with the mainland. All of these elements seriously undermined the representation of the city being 'sold' in the bid documents, that took for granted Trieste's strategic position within European infrastructural networks, assuming that the BIE would have bought into this storyline, no questions asked.

Trieste's 'European' Centrality

The idea that Trieste is destined to play a key role in the context of a redefined Central Europe and should be accordingly 'rescaled' as a regional capital, is more of an imagined scenario than a reality at the time of writing. It is a rather simplistic vision that derives from a set of widely accepted metaphors used to describe Trieste within such rhetoric: again, Trieste's natural geographical 'centrality', Trieste as the 'bridge' between East and West, Trieste as the 'door' to the East, and so on. This idea, presented by the Trieste Expo Challenge as one of the driving strengths of the candidature, is actually quite old and has been used and sometimes 'abused' by political actors of all persuasions over the last two decades, in order to inject optimism about the future of the city into the Triestini, as noted by historian Giampaolo Valdevit in the aftermath of the loss of the Expo (see Valdevit 2004). This idea became almost hegemonic in public discussions about the future of the city: already in June 1990, on the occasion of a conference organized by the provincial government launching the idea of an *offshore* financial centre, the local newspaper *Il Piccolo* proclaimed: 'La Trieste degli anni Novanta al centro della nuova Europa' [Trieste of the 1990s [will be] at the centre of a new Europe]. The aim, in this case, was to place Trieste 'at the heart' of a set of European financial and commercial flows, making it a new 'European hub', especially after the fall of the Berlin Wall (*Il Piccolo* 1990; *Il Messaggero del Lunedì* 1990; *Milano Finanza* 1990; *Il Sole 24 Ore* 1990). Nevertheless, in the decade or so that followed, among the plethora of new ideas and plans, nothing really significant was realized: not the Polis

14 For further information, see Luka Koper's website: <http://www.luka-kp.si>.

project previously cited, nor the offshore financial centre, nor the port, nor the much-vaunted trans-European transportation Corridor V.

Despite the long record of failures, the narrative adopted by the Trieste Expo Challenge during the bid drew precisely on those scenarios that had been long proclaimed by the local press and discussed in endless public debates. It is illustrative that a study published by the local Association of Entrepreneurs noted that, while such images were certainly important in 'pushing' some economic sectors, the metaphors adopted were of little use in planning any strategic development (Merzliak 2002). In particular, the imagined scenarios lost most of their appeal when it became patently clear that the very same kinds of positions and roles (related to ideas of centrality and 'bridging differences') were also being claimed by other cities in the area, from the Slovenian capital Ljubljana to the neighbouring Veneto region. A concrete example of the obstacles met by such imaginations in terms of infrastructural policies is given by the development of the Corridor V transport network. The Corridor was conceived as part of a set of Trans-European Networks (TEN) and aims to connect Lisbon in Portugal with Kiev in the Ukraine, via Turin and Trieste. Corridor V, in the last decade or so, came to represent for Trieste the concrete perspective of becoming one of the key infrastructural hubs of Europe. In practical terms, Corridor V would also strengthen the transborder connections between Slovenia and Italy, putting an end to the relative territorial isolation of Trieste, trapped between the Adriatic and the border. Some of the interventions envisioned as part of this initiative would include the construction of a high-speed and high-capacity railway link and the enlargement of the existing motorway. But the project is still in its planning phases as far as the railway link between Trieste Airport and the city centre is concerned, and only in the preliminary phases of the construction of a link between the Venice and Trieste airports. Considering the state of events at the time of writing, the projects are not likely to be completed for at least ten years.

Alongside the various technical problems associated with the fact that the territory through which the railway link would pass was particularly rough, another reason for the delay in the project was the high cost of the whole operation which had still not been covered in full at the time of the Expo bid. What is more, the project failed to consider (as did its environmental impact study) the possible overlap with other related infrastructural plans. For example, there is no mention of the project for the enlargement of the A4 motorway and a project termed *Penetrazione Nord*, an urban tunnel to be constructed in the very same area comprised by the Corridor (Predonzan 2004). Unsurprisingly, this disjuncture between competing plans and projects was not mentioned in the bidding documentation that presented the infrastructural plans for the region as all part of the Expo Master Plan (Trieste Expo Challenge 2004a, 168–9).[15]

15 This point is also stressed in a critical analysis of the bidding documents by the World Wildlife Fund (see WWF 2004).

Trieste's 'Urban' Centrality

As already noted earlier in this chapter, the project regarding the renovation of the Porto Vecchio envisaged by the Expo 2008 Master Plan was only the last in a long series of projects of this type. Had the bid been successful, the area was to become the international showcase for the hallmark event, thus creating a new functional centre – and even a new neighbourhood – for the city. However, here too the bidding documents and the numerous presentations organized by Trieste Expo Challenge neglected potential and existing conflicts and the overlapping competences, projects and stakeholders engaged in the renovation of the area.

For example, the last project to be presented by the Port Authority, under the aegis of Trieste Futura, began its process of legislative approval in 2001. The process was interrupted by an intervention by the Ministry of Culture which imposed new conditions on the renovation of some historical buildings. As a consequence, the project was modified again in 2002 by the Port Authority and the Municipality was asked to approve the changes, also in accordance with other planning requirements. An agreement between the Port Authority and the Municipality is required under Italian law in order to allow for an integrated approach between city planning and port development. Nonetheless, the Port Authority did not wait for this final approval and in 2003 opened the tender to assign different areas of the Porto Vecchio to public and private companies for redevelopment. As was to be expected, the whole procedure was frozen by the Regional Administrative Court (Tribunale Amministrativo Regionale, TAR).[16] For this reason, at the time of the presentation of the bidding documents and of the Master Plan for the Expo area in February 2004, it was still very uncertain which legal framework was required to be able to start the process of regeneration of the Porto Vecchio. In other words, the scenarios depicted in the proposal presented to the BIE, both for the exhibition and for the re-utilization of the Porto Vecchio after the event, were not based on concrete, legally actionable possibilities.

The political and legislative debates were rendered even more complex by the distinct set of fiscal privileges assigned to the port by the terms of the 1947 Peace Treaty, implicating also aspects of international law in the management of this piece of land. In particular, a certain degree of resistance towards the foreseen changes was manifested by some of the more conservative political and economic actors in the city, concerned about the preservation of the traditional maritime functions of this area. The most important object of controversy was whether the Port of Trieste should be subject to the same legal regime of all other Italian ports or whether it fell, rather, under the old International Treaty that stipulated a legal internationalization of the port area, removing

16 TAR ruling, Friuli Venezia Giulia, n. 285, 2004.

it from under Italian law.[17] The new uses for the Porto Vecchio envisioned by the plans for its role in a broader project of urban regeneration, it was argued, would mean a breach in international law (Querci 2001). Although the local coalition supporting this view had lost support in the years preceding the bid, and although also the political parties close to this conservative interpretation of the treaty expressed the need for a redevelopment of Porto Vecchio's maritime activities, the controversy remained, and proved an important factor in defining the future destination of the area. Notably, in the bidding documents this legal debate – which was becoming highly controversial in political terms – was not only omitted, but was also significantly underestimated in its potentially disruptive consequences for the whole project (see Maranzana 2004). The controversy was, in fact, not simply a technicality, but rather a symptom of the existence of several parallel and conflictual visions for the future of the Porto Vecchio. The main result of this conflict was, in practical terms, a stop to all the projects advanced in the previous decade for the regeneration of the Port which, *de facto*, undermined one of the key rationales for the Expo bid.

Networks, Place and (Not Much) Trust

The reason why it is so important to focus on the various representations of the supposed centrality of Trieste is that those different 'visions' reflect, in many ways, the variety of actors involved in the promotion of the candidature and the different scales on which the candidacy played out. What appears in the background of the project is an attempt at rearranging the structures of territorial governance, involving a complex mix of institutional levels directly or indirectly associated with the Expo bid. This spider web of affiliations and interests is mapped out in Figure 4.6 in an attempt to identify some of the major stakeholders and their relative links. In particular, we attempt to identify the key actors and the relationships explicitly associated with the formal process of the city's candidature: the BIE at the international level; the Italian government, including the diplomatic and political role played by the Ministry of Foreign Affairs at the national level; and a local coalition, represented by the Trieste Expo Challenge. This latter can also be envisioned as a distinct set of inter-scalar relations: it consisted, in fact, of a public-private partnership including the municipal, provincial and regional governments, together with the Port Authority. Furthermore, Figure 4.6 shows a process of scalar reconfiguration that involved, directly or indirectly, the city of Trieste as part of other, parallel, projects: the constitution of a Euroregion, the above mentioned Corridor V and, of course, the regeneration of Porto Vecchio.

17 The international regulations for the area of the Trieste Free Port are stipulated by the Peace Treaty signed in Paris on the 10 February, 1947, between Italy and the Allied Forces (see Querci 2001).

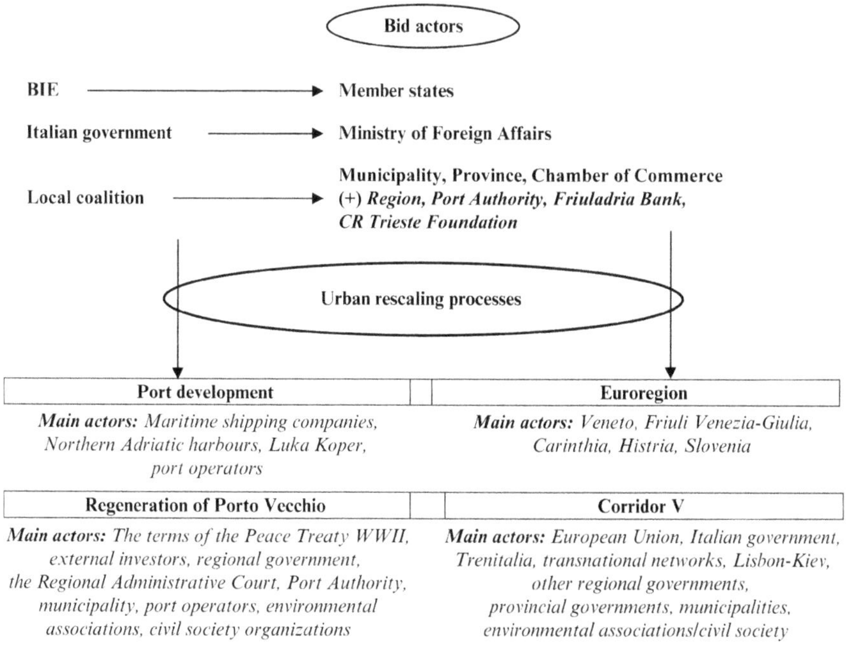

Figure 4.6 Institutional inter-scalar relations during the Expo 2008 bid

Notably, the link with international scientific institutions, so heavily emphasized during the bid and a key support to the central theme of the Expo, is absent from this scheme since, in our view, such a link was largely imagined: rather than being part of a broader strategy within the bidding process, it was limited to the mere celebration of the city's existing 'knowledge networks'. This is confirmed by the fact that in the plans for the re-utilization of the exhibition areas, no investment in scientific activities was envisaged. However, this complex network of relationships raises an interesting series of questions regarding the impact of such multi-scaled governance on the international positioning (and image) of the city and on the final result of the bid for the Expo.

In many ways, the chart in Figure 4.6 can be seen as a sort of mapping of the institutional attempt to mobilize a multi-scalar social capital for the city of Trieste. Our thesis here is that this mobilization failed to achieve its objectives because of a problematic gap between the image of Trieste (and its networks) presented in the international arena (and, in particular, to the BIE) and the actual degree of trust and collaboration among the institutions involved. This gap emerged very clearly in the final stages of the bidding process, paving the way to the exclusion of Trieste's candidature. The failure is also a perfect illustration, however, of the role of spatialized understandings of social capital and of how the functioning of a local network of trust can actually make the difference when faced with an event of this breadth.

It might be useful to return to an examination of the nature of the 'local coalition'. The different visions expressed and pursued by the different actors involved in the Trieste Expo Challenge – differences concerning not only specific parts of the project, but also the overall urban impact of the Expo – have already been explored. We can cite, therefore, a couple key examples. The first concerns the particularly vexed question of the regeneration of the Porto Vecchio: while both the Municipality and the Port Authority were supporting the Expo bid, they were fundamentally divided on how to interpret the legal framework for planning and port renovation. Other examples can be found in the numerous conflicts over the *politics of scale* to be adopted for the international promotion of Trieste. A concrete case of this latter concerns the projected role of Corridor V. In addition to a set of pre-existing technical and financial problems, this project overlapped with a number of competing plans concerning general mobility in the city, provoking several controversial public debates among local politicians and civil society representatives: the 'international' and the 'local' dimensions came into clear conflict here. Another area of conflict was the future of the Port of Trieste as a whole: on the one hand, there was institutional and public support for the implementation of a stronger trans-border integration of the Northern Adriatic port system, through intensified co-operation with the ports of Koper (Slovenia) and Rjieka (Croatia); on the other, equally strong opposition to this plan was expressed by other stakeholders, concerned with the maintenance of the status quo and advocating a stronger integration with the national port system instead.

Given such a controversial political context, it should not come as a surprise that the proposal advanced by Trieste Expo Challenge was marked by a conspicuous absence of a shared and consistent vision for the future of the Porto Vecchio and the city as a whole. We can interpret this as due to a 'scarcity' of urban social capital and local networks of trust, often seen as the key starting point for the successful organization of an event of the magnitude of the Expo. What is more, the agency in charge of managing the bid process essentially limited its task to the collection and collation – within a single document – of basically all the urban projects 'on the table' at the moment of the submission of the bid, a hodge-podge of projects that clearly revealed overlapping and conflicting plans. As a consequence, the final proposal was a rather confused document, lacking clear method and weak in terms of its contents. In addition to this, to many actors involved in the process, both the organization and the projected impact of the bid appeared to be conceived within a very narrow 'geographical horizon'. A clear indicator of this was the communication strategy adopted, which did not go beyond the regional level, with only few interventions in the national press and a clear lack of engagement with the international media.[18]

What is more, the relationships between the local coalitions and the other scales involved in the bidding process remained highly problematic. A good example of this is the way in which the national government was involved in the entire process.[19] The Italian government gave its official support to Trieste's bid in the spring of 2003, with the candidature letter forwarded by the then Minister of Foreign Affairs, Franco Frattini, to the Secretary General of the BIE, and with the constitution of a coordinating committee for Expo 2008, including representatives of the Ministry and Trieste Expo Challenge. At that point, the focus of the local coalition was mainly on obtaining public funding: €862 million were in fact promised, €560 million to come from the national government and the rest to be raised from the re-utilization of the area after the end of the Expo (Morelli 2004). The funding question explains quite well the kind of urban regime envisaged by the heterogeneous coalition that managed the bid process. The strong internal disagreements over essentially all major strategic points of the bid and its potential impacts cannot possibly be interpreted through the standard 'growth coalition' framework described by Logan and Molotch (see 1976; 1987).[20] Rather, it is probably more correct to envision the bid as a *grant coalition*, similar to that described by Cochrane, Peck and Tickell (1996) in the case of Manchester's candidacy for the Olympic Games. Such an interpretation, indeed, allows us to highlight another key characteristic of local politics in Trieste: contrary to arguments positing the progressive loss of state power (highlighted by many analyses of urban growth coalitions), the Trieste bid revealed that local actors often regard their relationship with the central state as the most important factor in the future development of the city.

Moreover, the bid also revealed the lack of a strong and cohesive local milieu able to act as 'collective actor' for developmental goals. Rather, what took shape was an alliance of competing interests, aimed at obtaining funds from the state. The political debate that followed the bid's failure confirmed, in many ways, this attitude, assigning the bulk of the responsibility for the negative outcome to the national government. As highlighted by the local press, instead of focusing on the mistakes made at the city level, the debate was centred on the role of

18 Our interview with Fabio Carini (20 July 2004), Press Officer for Trieste Expo Challenge, confirmed that the communication strategy was targeted specifically for a local audience. On the national level, between February and November 2004, only one article on the *Corriere della Sera* daily newspaper and the weekly *Panorama* magazine highlighted the bid. International media were not considered as a target.

19 The conflictual relation between local and national politicians was particularly evident in a certain phase of the bid (for further analysis, see Benvenuti 2004).

20 The 'urban regime theory' was formulated in the 1980s, taking into consideration a plurality of actors and various possible combinations of strategic alliances at the city level. See, among others, Fainstein and Fainstein 1986; Elkin 1987; Stone 1989).

Italy on the international scene, at the same time neglecting any analysis of the proposals drafted in Trieste.[21]

Assigning the bulk of the responsibility for the bid's failure to the national (central) state was, however, also in some sense a reflection of the city's distinct historical-political context: Triestine politics had long been characterized by relations of strong dependence with the 'patron state' – for many decades Vienna and, in the post-war period, Rome (see Minca in this volume; Rumpler 2003). Even more important, however, was the widely shared sense of historical loss felt by the Triestini after WWII, a loss that the Italian state was expected to in some way compensate, paying back the enormous sacrifices suffered by Trieste and its people (see Minca forthcoming; Valdevit 2004).

More generally, the Trieste bid highlights two crucial elements concerning, on the one hand, the analysis of the impact of major events within the international competition among cities, and on the other, the ways in which social capital, or its absence, can affect local governance and the capacity of a city to attract investments and 'fresh' cultural capital. Firstly, the case of Trieste demonstrates that in order to create a solid international image of urban 'centrality', the deployment of marketing strategies and local/regional promotion are not enough. In many ways, Trieste's candidature suffered from a long series of pre-existing political and infrastructural problems and was therefore unable to present a coherent vision (and set of policies) convincing enough for the delegates of BIE to assign the Expo to the city. Our investigation noted that, beyond the catchy slogans and apparent unity of purpose, the bidding process was torn by tensions between different actors and their divergent scales (and modes) of intervention – tensions evident even *within* the bid coalition. Conflict, rather than cooperation, marked the interaction between key actors and stakeholders. What is more, such conflictual relations at the local scale were further compounded by the equally problematic role of the central state: in particular, its projected role in funding new urban development and promoting the internationalization of the city.

Other aspects of the process, of course, also deserve further analysis. One is the role of local civil society in the bidding process; another is the question of the potential social and cultural impact of Trieste's candidature. It is difficult to speculate on these points, however, since following the failure of the bid, the project was simply set aside; it remained in the realm of political discourse and never went beyond the initial declarations and expectations. Nevertheless, even a superficial analysis of the approach adopted by Trieste Expo Challenge reveals the lack of any real involvement on the part of the city's inhabitants in

21 This 'nationalization' of the debate can be resumed by the irony of a short article on the local daily *Il Piccolo* on 20 December 2004: 'Manifesti a tutto muro "Berlusconi ha sostenuto l'Expo 2008 a Trieste". Abbiamo visto il risultato. Grazie di tutto, continui così' [The posters across the walls read: 'Berlusconi supported the Expo 2008 in Trieste'. Great, we saw the results. Thanks].

the debates over the future of their city. Popular participation was presumably gauged by a survey (whose results were included in the bidding documents), suggesting that the candidature was supported by 93 per cent of the Triestini. The active involvement of the citizens was to be channelled through an association explicitly created for this purpose: ProTrieste Expo. The actions of the latter, however, were limited to the organization of public displays of support for the activities of Trieste Expo Challenge, a 'support' that did not envision a proactive role for the local population.[22]

Conclusions

The failed Expo bid left an important mark on Trieste. The quiet shock of the first few days was followed by a great deal of debate over the future of the city, within which it soon became clear that the absence of a consolidated set of local networks of trust and of a recent tradition of 'working togetherness', capable of mobilizing the existing (in potential, at least) urban social capital, had been a fundamental factor in the failure of the bid. The misinterpretation of the necessarily multi-scalar and cosmopolitan dimension of any such bid on the part of the promoters, coupled with an objective impossibility (at least at that moment in time) of galvanizing the support of a cohesive and single minded urban elite, was seen as the unfortunate result of Trieste's incapacity in mobilizing the sort of 'local' social capital on which institutions like the BIE normally rely for the successful organization of an international event of this breadth.

The presence of underlying conflicts over competences and jurisdictions and, more generally, a marked incapacity to pursue 'together' the city's general interest (that, presumably, corresponded to the success of the bid) on the part of many institutional and non-institutional actors in Trieste resulted in a number of important shortcomings. Firstly, it produced a very problematic gap between the image of Trieste as promoted by the bid – and the actual political and economic situation of the Adriatic city. Secondly, there was an excessive reliance on the part of the promoters on the support of the Italian state – certainly an important element of any such bid, but in Trieste's case one that cast doubt on the capacity of the city itself to manage the event. Thirdly, we can note a visible absence of an immaterial but substantial network of trust that should have been mobilized by the urban elite around the candidature in order to achieve an important common objective for the city. A final, and equally problematic

22 The lack of involvement by local civil society groups was decried by the local section of the World Wildlife Fund (WWF) to the authorities of the BIE. This episode came as a reaction to the fact that the bidding documents explicitly referred to a process of 'wide consultation with environmental organizations' that did not take place (for further information, see http://ambiente.triesteincontra.it/wwf).

shortcoming, was the extremely modest and relatively passive involvement of the local population in the initiative – again, an involvement that would normally be considered a crucial element of support in any competition of this sort.

It appears that Trieste has begun to learn from its mistakes, however. A number of important shifts have, in fact, significantly transformed the prospects of the city in the immediate post-bid period. 2007 brought two key events – the definitive disappearance of the border with Slovenia, and plans for the constitution of a new Euroregion,[23] with Trieste nominated as its putative capital. In both instances, those very same city institutions that failed to collaborate during the bidding process – for instance, the Municipality, the Region and the Port Authority – this time around seized the new opportunities and began to openly collaborate, mobilizing new formal and informal networks of trust and getting these directly involved in shaping of the future of the city. By leaving behind the long list of unaccomplished projects and failed collaborations, the Triestine urban elite is now trying to 'work together' in order to valorize an urban social capital that, despite a past institutional culture of conflict, appears to be emerging, when the general interest of the city is at stake. But, as we well know, social capital is an immaterial thing and works only when 'activated' in order to pursue a specific project in a specific place: the future will tell if this is just another (mid-summer) dream – or if Trieste's famed cosmopolitan culture of trust is back again.

Bibliography

Ara, A. and Magris, C. (1982), *Trieste, un'identità di frontiera* (Turin: Einaudi).

Arrighi, G. (ed.) (1985), *Semiperipheral Development: The Politics of Southern Europe in the Twentieth Century* (London: Sage).

Autorità Portuale di Trieste (2001), *Variante Generale al Piano Regolatore Portuale per l'area del Punto Franco Vecchio e Barcola* (Trieste, art. 1, Introduction).

Barduzzi, O. (2004), 'Il piano strategico, strumento per le politiche di efficienza istituzionale e per migliorare la capacità competitiva della città: l'esperienza di Trieste', in V. Fedeli and F. Gastaldi (eds).

Basso, S. (2004), 'Trieste: regesto degli strumenti urbanistici 1991–2001', *Urbanistica* 123, 60.

Beck, U. (2006), *The Cosmopolitan Vision* (Cambridge: Polity Press).

Billinge, M. (1993), 'Trading history, reclaiming the past: the Crystal Palace as icon', in G. Kearns and C. Philo (eds).

Boeri, S. (ed.) (2001), *Progetto per il recupero del Porto Vecchio di Trieste* (Trieste: Autorità Portuale di Trieste).

23 Including the region of Friuli Venezia Giulia, the Veneto, Carinzia and parts of Slovenia and Croatia.

Boileau, A. and Cimiotti, M. (eds) (2004), *Popolazione, Occupazione, Pensioni, Abitazioni, Famiglie a rischio. Quarto rapporto* (Trieste: Osservatorio Provinciale sulle Politiche Sociali e le Nuove Povertà – Provincia di Trieste).

Borruso, G. (1999), *Riflessioni intorno a un sogno di mezza estate da tradurre in realtà* (Trieste: TriestExpo).

Borruso, G. (2004), 'Trieste Futura, TriestExpo, Triesteidea', *Urbanistica* 123, 80–83.

Camera di Commercio di Trieste (ed.) (2004), *Atti della conferenza internazionale: Portualità e Trasporti nell'Europa allargata* (Trieste: ISDE).

Carapetian, M. and Caputo, F. (eds) (2000), *Feasibility Plan Trieste Expo* (Venezia: C. & C. Architettura e Ingegneria).

Caroli, A., Cociancich, M. and Della Puppa, M. (2003), *Il porto di Trieste e la sua localizzazione geostrategica nel bacino del Mediterraneo* (Trieste: Edizioni Italo Svevo).

Chalkey, B. and Essex, S. (1999), 'Urban development through hosting international events: A history of the Olympic Games', *Planning Perspectives* 14, 369–94.

Cochrane, A., Peck, J. and Tickell, A. (1996), 'Manchester plays games: Exploring the local politics of globalisation', *Urban Studies* 33:8, 1319–36.

Collotti, E. (1974), *Il Litorale Adriatico nel nuovo ordine europeo 1943–1945* (Milan: Vangelista).

Colombino, A. (forthcoming), 'Multiculturalism and time in Trieste: Place-marketing images and residents' perceptions of a multicultural city', *Social and Cultural Geography*.

Colombino, A. and Minca, C. (2005), 'Spettacolo e gentrificazione sociale a Trieste', in C. Minca (ed.).

De Spuches, G. (2002), 'La fantasmagoria del moderno, Esposizioni Universali e metropoli', *Bollettino della Società Geografica Italiana* 12:7, 783–94.

Di Biagi, P. (2004), 'Trieste. Laboratorio di politiche, piani e progetti', *Urbanistica* 123, 54–6.

Elkin, S. (1987), *City and Regime in the American Republic* (Chicago: University of Chicago Press).

Fainstein, N. and Fainstein, S. (1986), 'Regime strategies, communal resistance and economic forces', in S. Fainstain et al. (eds).

Fainstein, S., Hill, R.C., Judd, D. and Smith, M. (eds) (1986), *Restructuring the City: The Political Economy of Urban Redevelopment* (New York: Longman).

Finzi, R., Magris, C. and Miccoli, G. (eds) (2002), *Storia d'Italia. Le regioni. Il Friuli-Venezia Giulia* (Turin: Einaudi).

Finzi, R., Panariti, L. and Panjiek, G. (eds) (2003), *Storia economica e sociale di Trieste. Volume II. La città dei traffici 1719–1918* (Trieste: Lint).

Guagnini, E. (2003), 'Trieste: ponte tra culture, postazione di confine', in R. Finzi et al. (eds).

Kearns, G. and Philo, C. (eds) (1993), *Selling Places: The City as Cultural Capital, Past and Present* (Oxford: Pergamon Press).
Lodato, F. (2001), 'Storicità del porto franco di Trieste', in F.A. Querci et al. (eds).
Logan, J. and Molotch, H. (1987), *Urban Fortunes: The Political Economy of Place* (Berkeley, CA: University of California Press).
Marin, A. (2004), '1991–2001: dal piano regolatore al piano strategico', *Urbanistica* 123, 59–70.
Merzliak, S. (ed.) (2002), *Ripartire dai paradigmi industriali, strategie di rete* (Trieste: Associazione degli Industriali della Provincia di Trieste).
Merzliak, S. (2003), *Ricerca e innovazione, infrastrutture e porto: la relazionalità per lo sviluppo* (Trieste: Associazione degli Industriali della Provincia di Trieste).
Millo, A. (2003), 'Un porto tra centro e periferia (1861–1918)', in R. Finzi et al. (eds).
Minca, C. (1995), 'Urban waterfront evolution: The case of Trieste', *Geography* 80, 225–34.
Minca, C. (ed.) (2005), *Lo spettacolo della città* (Padova: Cedam).
Minca, C. (forthcoming) 'Trieste nazione and its geographies of absence', *Social and Cultural Geography*.
Molotch, H. (1976), 'The city as growth machine', *American Journal of Sociology* 82:2, 309–55.
Pichierri, P. and Pugliese, A. (2004), *L'Expo di Trieste* (Trieste: Iniziative Culturali).
Predonzan D. (2004), *Il progetto della tratta AV/AC Venezia (Ronchi dei Legionari)* (Trieste: WWF).
Pritchard, B. (2004), 'Cooperazione fra i porti e prospettive per lo short-sea shipping in Adriatico', in Camera di Commercio di Trieste (ed.).
Querci, F.A. (2001), 'Il porto di Trieste come territorio internazionale', in F.A. Querci et al. (eds).
Querci, F.A. (2001), 'Limiti di giurisdizione nel Porto Franco di Trieste', in F.A. Querci et al. (eds).
Querci, F.A., Trampus, F. and Lodato, F. (eds) (2001), *Internazionalità e storicità del porto franco di Trieste* (Trieste: EUT).
Rumpler, H. (2003), 'Economia e potere politico. Il ruolo di Trieste nella politica di sviluppo economico di Vienna', in R. Finzi et al. (eds).
RUR/Censis (ed.) (2004), *Rapporto Municipium 2004* (Roma: RUR/Censis).
Starc, W. (2004), 'Il territorio della provincia di Trieste nella nuova Europa', *Urbanistica* 123, 90–92.
Stone, C. (1989), *Regime Politics* (Lawrence: University Press of Kansas).
Tange, K. and Urtec (1971), *Linee guida per lo sviluppo dell'area di Trieste* (Trieste: Studio Tange).
Trieste Expo Challenge (2004a), *Documents for the Inquiry Mission of the Bureau International des Expositions* (Trieste: Trieste Expo Challenge).

Trieste Expo Challenge (2004b), *Trieste Expo 2008 News* (Trieste: Trieste Expo Challenge).
Trieste Expo Challenge (2004c), *Opuscolo Informativo* (Trieste: Trieste Expo Challenge).
Trieste Expo Challenge (2004d), *Slides from the Official Presentation of the Candidature at the BIE, Paris*, June 2004 (Trieste: Trieste Expo Challenge).
Trieste Futura (1998a), *Opuscolo Informativo* (Trieste: Trieste Futura).
Trieste Futura (1998b), *Sessioni di lavoro, 19 dicembre 1998* (Trieste: Trieste Futura).
Valdevit, G. (1986), *La questione di Trieste 1941–1954. Politica internazionale e contesto locale* (Milan: Franco Angeli).
Valdevit, G. (2004), *Trieste. Storia di una periferia insicura* (Milan: Bruno Mondatori).
Waitt, G. (1999), 'Playing games with Sydney: Marketing Sydney for the 2000 Olympics', *Urban Studies* 36:7, 1055–77.
Wallerstein, I. (1985), 'The relevance of the concept of semiperiphery to Southern Europe', in G. Arrighi (ed.).

Press Articles

Baldassi, F. (2004), 'Brunetta: faremo l'Expo anche senza Expo', *Il Piccolo*, 20 December.
Benvenuti, D. (2004), 'Paoletti: non abbiamo perso la guerra', *Il Piccolo*, 20 December.
Benvenuti, D. (2004), 'Expo, nessuno è uscito indenne', *Il Gazzettino*, 20 December.
Contessa M. (2004a), 'Navi in fuga dal porto, ecco i perché della paralisi', *Il Piccolo*, 21 July.
Contessa M. (2004b), 'Ma il 2004 si era aperto in forte ripresa', *Il Piccolo*, 21 July.
E.R. (1990), 'Un coro di sì a Trieste off-shore per banche e commercio', *Il Sole 24 Ore*, 20 June.
Maranzana, S. (2004), 'Parco sul mare, subito il primo vertice', *Il Piccolo*, 18 December.
Il Messaggero del Lunedì (1990), 'Commercio e finanza: ruolo internazionale', 18 June.
Milano Finanza (1990), 'Trieste vuole diventare boutique finanziaria', 20 June.
Morelli R. (2004), 'Trieste vuol essere capitale della scienza', *Il Corriere della Sera*, 19 April.
Maranzana, S. (2004), 'E ora via con il Porto Vecchio, ma senza ambiguità', *Il Piccolo*, 19 December.

Il Piccolo (1990), 'La Trieste degli anni Novanta al centro della nuova Europa', 13 June.

Il Piccolo (1998), 'Expo universale a Trieste? Si può', 15 December.

Taino, D. (2004), 'Battuti dagli spagnoli: non facciamo sistema', *Corriere della Sera*, 18 December.

Valdevit, G. (2004), 'Bilancio del libro dei sogni', *Il Piccolo*, 19 December.

WWW (2004), 'Dossier per la candidatura di Trieste all'Expo 2008', Osservazioni, Trieste.

Chapter 5
Trust in Translation: The Role of Place and Vision in Building the eTampere Programme Network

Timo Poutiainen and Jouni Häkli

Introduction: 'Let there be light!'

After fruitless experimenting with platinum and other metal filaments on 22 October 1879, Thomas Alva Edison finally discovered the carbon material that would enable commercially successful production of electric light, and thus, seal his fame as one of the most successful inventors in the history of modern technology. This groundbreaking discovery was made in Edison's industrial research laboratory in Menlo Park, New Jersey – arguably the first 'innovative environment' of its kind (Pretzer 2002). In Menlo Park chemists and engineers carried out joint research and development work under the direction of Edison who at the end of his career held more than a thousand US patents in his name.

One of the Menlo Park engineers in the late 1870s was Carl von Nottbeck, the oldest son of industrialist Wilhelm von Nottbeck who owned the Finlayson cotton mill in Tampere, Finland (Figure 5.1). Carl von Nottbeck studied under the guidance of Edison and became one of his apprentices in electric engineering. When returning to Tampere from the United States in 1881, this young engineer brought with him Istvan von Fodor, a skilled Hungarian-born engineer from the Edison Electric Illuminating Company. With Fodor's help, Nottbeck intended to set up electric lighting in the spinning hall of the Finlayson factory.

While pleased about the prospect of increased fire safety and labor efficiency, the industrialist Wilhelm von Nottbeck was somewhat hesitant about his son's plan to replace gas lighting with electric light. However, successful testing on 15 March 1882 of the two 110 volt Edison direct-current dynamos and the attached system of 150 light bulbs swept away all remaining doubts. Celebrating the triumphant day, Nottbeck and Fodor sent Edison a telegram: 'Lighting installed at 61 degrees northern latitude. Complete success' (Myllyntaus 1991, 302). Soon the use of electric light was expanded in the factory and a few years

Figure 5.1 The location of Tampere, Jyväskylä and Haparanda-Tornio

later an initiative was made to bring electric lights to the streets of Tampere. The growing demand for electric power, again, prompted the establishment of Finland's first public electricity works in Tampere in 1888 (Myllyntaus 1991).

Through this chain of events, Tampere has won a remarkable primacy: she was the first town in the Northern Europe to introduce electric lighting and, remarkably, fifth in the whole of Europe, along with cities such as London, Paris, Milan and Strasbourg. This is a notable achievement for a town located on the northern European periphery and with a population of less than 15,000 inhabitants at that time. However, what makes this history significant is less its peculiarity than the way in which it has given rise to a very specific image and narrative of Tampere. Starting from the manner in which the electrification of the town was noted in local press, praising the transfer from 'wretched oil lamps' to wondrous 'electric sunlight', all the way to present day place marketing leaflets boasting Tampere's pathway to industrial success, the story has not lost its appeal.

The circumstances that led to the early adoption of electric light in Tampere would make a prime case for tracing the complex networks of actors compressing time-space up to a point where a peripheral city at the 61st parallel, located in a poor and sparsely populated country, suddenly got connected with the very core of technological development. However, in this chapter we shall attempt to read this history in broader terms. What we are interested in is the way in which the narrative of Tampere as a city on the leading edge of technological progress, deriving from the late nineteenth century, was revitalized as part of the official vocabulary revolving around a contemporary development scheme called the *eTampere programme*. More precisely, we wish to assess the manner in which the progressive identity of Tampere was successfully inscribed into particular visions and goals that materialized in the eTampere programme and helped shape its contents and contours.

We set out to examine the eTampere network formation from the viewpoint of social capital and trust, with emphasis on the planning and early implementation of this ambitious development project. We will also employ aspects of actor-network theory (Callon 1986; 2001; Latour 1993; 1999) so as to better grasp the dynamics of networking, and to provide an illustrative example concerning the enactment of social capital in network-like organizations. Our main focus is on the examination of the role of the eTampere programme vision and goals as narrated in various programme documents as well as by the participants themselves. By programme goals and visions we mean the formal and informal aspirations of eTampere, conveyed in both written and spoken form. The *vision* is a general view of the hoped-for future to be reached through strategy implementation and fulfilling the stated *goals*. These goals are place-specific as they reflect and appropriate narratives concerning the history of Tampere and the local way of 'getting organized and doing things'. We are particularly interested in the degree to which these stories can be interpreted as a form of cognitive social capital that builds a sense of togetherness among heterogeneous actors and directs the processes of recruiting and engaging new members into the programme network.

We begin by outlining the eTampere programme and the conditions in which it was launched. We then seek conceptual tools for understanding this development by foregrounding the idea of stories as a 'glue' in social interaction, and by relating the discussion to the concept of social capital with particular emphasis on its cognitive dimension. We then turn to actor-network theory for analysing and interpreting network formation. By paying attention to stories as expressed in eTampere documents from the viewpoint of 'translation' (Callon 1986), we examine the concrete formation of the eTampere network. Finally, we conclude by discussing what we can learn about the dynamics of social capital in the case of an ambitious development programme and the networks that were built to implement it.

Towards World Leadership: Launching the eTampere Programme

> Tampere is widely known for its bold and successful ventures. In many issues, Tampere has been one of the European leaders and even world leaders. An early example is the electrical lighting introduced at the Finlayson factory and more recently the Communicator [mobile phone] developed by Nokia. … The eTampere programme, a five-year development project costing 130 million euros, will open the new millennium. Its general objective is to make Tampere a global leader in the research, development and application of issues related to the Information Society. … Tampere will now take a giant leap into the future – bold and dedicated. (Jarmo Rantanen, Mayor of Tampere, see eTampere Programme Plan 2000)

At the turn of the millennium, enthusiasm concerning the information society was at its apex. Visions all over the industrialized world were monumental and the possibilities of the ICT sector seemed almost boundless. Such optimism reigned also in Tampere, the third largest city in Finland, where the atmosphere and the conditions seemed favourable for 'the next big thing' to emerge. As expressed in the eTampere programme:

> There was a firm and widely held belief in Tampere that the city had excellent prerequisites to become world leader in the development of the knowledge society, and it was necessary to take action to reach that position. (eTampere website 2007)

In other words, there was a perceived need to update the city's traditional strengths – education, communication and culture – to meet the challenges of the information society and the new millennium. The enthusiasm quickly turned into action, and preparations for the eTampere programme started in spring 2000.

The eTampere initiative was launched as a response to the realization that the information and communication technology sector in Tampere needed to be strengthened to ensure that technological development would not bypass the city. The initiative was brainstormed by a number of key stakeholders in ICT research and development and it matched up well with the city's urban renewal policies concerning the old industrial premises in the Finlayson area. Backed up by the eEurope initiative, eTampere was also aimed at enhancing the attractiveness and competitiveness of the Tampere region that was still seen as being labeled by its heavy industrial image (Lähteenmäki-Smith and Steineke 2003).

Designed so as to find and build on new approaches to cooperation and networking, the eTampere programme followed explicitly guidelines adopted from theories of network society (Castells 1996), creative class and city (Florida 2002), and innovative milieu (Camagni 1991). Among the dominant themes in the programme scheme were the significance of tacit knowledge, the rise of the global information society, and the controlling of different flows such as capital and information. The focus was on cross-organizational and multidisciplinary networks with both their weak and strong ties. Assuming that the most important things would happen in and through these networks, their dynamics and membership in them were seen as critical sources of capacity building (eTampere Programme Plan 2000).

The general objective of the eTampere programme was no less than to make Tampere a global leader in the research, development and application of products and services related to the information society. There were three themes within the programme. The first was to develop network services that ease citizens' daily lives and are accessible to everyone. Second, the programme aimed to strengthen the basis of expertise in research and education. The third objective was to create new innovative businesses. The more specific objectives included, inter alia, generating twenty new, globally competitive eEconomy companies with a total personnel of 1,500 and a turnover of €250 million, and reducing the overall unemployment rate of the city by five percentage points (eTampere Programme Plan 2000). The eTampere programme was spectacularly announced in the regional newspaper *Aamulehti* of 14 October 2000, and the Programme Plan was ready in December 2000. The programme duration was set for five years (2001–2005), and the total fiscal volume of projects within the programme and its six sub-programmes was estimated to reach €130 million.

Strongly linked to the strategic planning of Tampere, the eTampere programme was from the outset seen as a natural extension of the city's continuum of large and ambitious projects (see for instance Särkänniemi amusement park, Näsinneula tower, and the Tampere Hall concert and congress center), as well as numerous world-first technical innovations (for example, NMT, GSM and Internet calls, Timberjack walking forest harvester – see eTampere Programme Plan 2000). However, this time – in the spirit of the information society and the emergent globalization – the goals of the programme were less concrete and focused. Instead of 'building walls and activities tied to one location', the programme concentrated more on emerging extensive cross-disciplinary and cross-organizational cooperation and networking between business life, government and universities, so as to accumulate and increase the local intellectual capital (eTampere Programme Plan 2000). This challenge played a dual role in the implementation of the programme: networking was a major objective in the programme, as well as the most important means to reach its goals. But where did those goals emanate from?

While there arguably is a continuum between the industrial history of Tampere and the vision behind the eTampere programme, its grandiose tone and goals nevertheless remain somewhat puzzling. Tampere was founded as a town in 1779 by a royal decree of the King of Sweden Gustav III on the bank of the Tammerkoski rapids. By the mid-nineteenth century the town had gained only a modest population of 5,500 inhabitants. However, during the latter part of the nineteenth century Tampere grew into a major market place and industrial centre both on a Finnish and Scandinavian scale (Figure 5.2). By the end of the century, Tampere had become the third largest town in Finland and the most important industrial centre, being home to roughly half of the country's industrial workers. This earned Tampere the epithet of the 'Manchester of Finland', an exaggerated designation for a city of 36,000 inhabitants, popularized by writer and journalist Zacharias Topelius in the late nineteenth century.

Figure 5.2 Renewed Finlayson factory buildings by the Tammerkoski rapids (photo by Jouni Hakli)

The textile industry was the most important field for a long time, and metal and mechanical engineering industries grew rapidly before World War II. As in most Western countries, the traditional, labour-intensive heavy industries started to decline in the 1960s, but still today Tampere evinces a substantial industrial activity. However, significant steps towards a knowledge economy and information society were also taken in the 1960s as two universities were founded in the city (the University of Tampere and the Tampere University of Technology). They have both grown into major universities in Finland, providing higher education to a total of 27,500 students (Kostiainen and Sotarauta 2003, 419–26).

Tampere, with a population of more than 200,000 the most important urban centre in Finland after Helsinki, has always been proud of her industrial roots. However, the industrial path of the city does not self-evidently lead to the adoption of a massive development programme with a budget in the range of 10 per cent of the city's total annual expenditure. While the city's own direct investment in the programme was just over €20 million, the €130 million total budget of eTampere is indicative of the grand scale of this development effort (Niinikoski et al. 2006). To account for the success of the city in launching a development programme of this magnitude, it is useful to turn to a discussion of narratives and stories as a force that helps in mobilizing heterogeneous actors to aspire for a shared vision and common goals.

Actors, Networks and 'Storied' Social Capital

Innovative and innovation-producing networks and organizations, such as the eTampere programme, typically bring together heterogeneous actors with dissimilar disciplinary backgrounds and mindsets (Granovetter 1992; Florida 2002). While the resulting tensions may be an enrichment and source of dynamism, viable communication as an essential part of social exchange between actors also requires common and shared elements. This is where shared language and stories turn out as salient points binding actors together (see Inkpen and Tsang 2005, 159).

In fact, stories such as organizational visions and narratives of local history may play a key role in organizing concerted social action (Boje 1999; Czarniawska 1997; 1998; Porcaro and Minca in this volume). Because stories usually have a plot that involves meaningful people to identify with, they bind people together more effectively than official, often neutral, pronouncements (Cohen and Prusak 2001). Stories tend to foster solidarity by creating a sense of membership and belonging. They can also be an important resource for cooperation as they spur collaborative effort by telling actors why they are working together, who they collectively are, and what they may become in the future (ibid., 119–25). At best, stories help to tie together individual and organizational identities and thus strengthen emerging networks (Gardner 1995; Mishler 1999).

On top of shared stories, social capital may influence significantly the composition and functionality of collaborative networks. In the 'Bourdieuan' understanding adopted here social capital refers to the assets that networked actors may mobilize through their contacts and connections (Bourdieu 1986; Burt 1992). Hence, social interaction is both a precondition for the development of dense social capital, and a means of enacting it (Bourdieu 1986). Because social capital is tightly connected to interaction and social exchange, factors that shape the evolution of social relationships also have an effect on the development of social capital. According to Nahapiet and Ghoshal (1998, 256), those factors include time, interdependence, and network closure. Time is important in the sense that establishing trustful relationships depends on the durability and continuity of social interaction. Interdependence, again, is significant because social capital tends to erode when actors become less dependent upon each other (Coleman 1990). Network closure intensifies social interaction and thus fosters the development of dense social relationships (Nahapiet and Ghoshal 1998).

According to Nahapiet and Ghoshal (ibid., 243) social capital has structural, cognitive and relational dimensions (Figure 5.3). In what follows, we concentrate mostly on the cognitive dimension, which we approach in terms of organizational stories and myths as potential factors helping organizations to achieve their goals (Cohen and Prusak 2001). Shared stories and common language facilitate more effective knowledge transfer, and build understanding between participants. They contribute to evoking tighter relationships between network members and thereby foster the emergence of shared visions (Tsai and Ghoshal 1998; Inkpen and Tsang 2005).

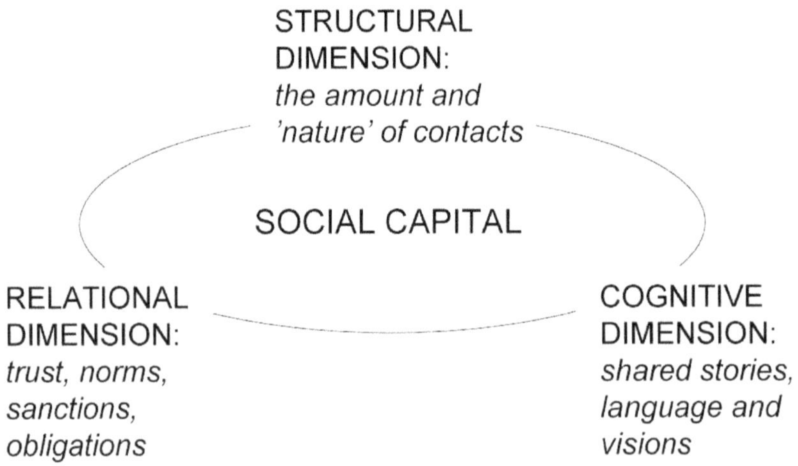

Figure 5.3 The dimensions of social capital (elaborated from Nahapiet and Ghoshal 1998)

To theoretically relate stories, interaction and networks and to empirically explore the case of the eTampere programme, we apply to the analysis conceptual tools from actor-network theory (Callon 2001). We adopt its 'weak version' that views actor-networks as both social (human) and material (non-human), but not in equal measures because it is social relations that direct the shaping of the networks (Murdoch 1998; Castree 2002; Whatmore 2006). According to Hacking (1999), social actors are 'interactive' while material actors are 'indifferent'. Hence, social actors are conscious about their position in the structure and environment around them. They use language, communicate, interact, and thus are aware of what is said and thought about them or done to them (Castree 2002, 31–2). This gives social actors the ability to learn new things, actively react to changes, and thus continuously negotiate concerning their network position (Bourdieu 1986).

From the point of view of this chapter, the most useful analytical concepts in actor-network theory are intermediary, translation and black box (Callon 1986; Latour 1999). The basic constitutive element of a network is a relationship between two actors and an intermediary. An 'intermediary' is 'anything passing between actors which defines the relationship between them' in the course of relatively stable transactions (Callon 1991, 134; Callon and Law 1992, 25). Intermediaries perform communicative relationships between actors and thus enable them to convey their intentions to each other. Michel Callon (1991, 135–40) has distinguished four types of intermediaries. The first type is texts, or literary inscriptions, such as contracts, strategy documents, books and reports. The second category comprises of technical artifacts and the third, human beings with all their technical and social skills and know-how. The fourth type is money in its various forms. In the following, we will pay specific attention to the role of eTampere programme documents as intermediaries encapsulating the programme vision and narratives of local history.

The actor-network takes its shape when actors and intermediaries describe the network by identifying and specifying other relevant actors, groups and intermediaries. These relationships are formed through the process of 'translation', wherein an actor aspires to readdress the interests of potential allies to parallel the interests of the actor-network, as well as to establish its own authoritative position to speak on behalf of allies (Åkerman 2006, 38). Translation is carried out in four stages as outlined by Callon (1986). Everything starts from *problematization*, where the primary actor formulates the manuscript for translation and defines the identities and interests of other actors that are consistent with its own interests. Problematization is followed by *interessement*, a stage where the primary actor tries to attract the other actors' interest and to negotiate their terms of involvement. The primary actor must convince the other actors that the roles defined for them are acceptable, and to establish the alliances. The last two stages of the translation process are fused together. In the stage of *enrolment*, actors accept the roles that have been defined for them during interessement. The final stage, *mobilization* of allies, follows if the previous stages are successful. The main concern regarding this final stage is whether actors act according to their 'agreement' (Palmroth 2004, 22).

Finally, durability in actor-network relations is enhanced by what Latour (1987, 1–3) calls 'black boxes'. The concept refers to the establishment of certain basic assumptions in a way that makes them immune from criticism. A black box, therefore, is any reality or claim that, no matter how complex or contested its history has been, has come to reach the degree of stability where it can be taken for granted. With the help of black-boxing, it is possible to build argumentation and knowledge and establish and stabilize network relationships. In the case of eTampere we will look particularly at how the programme vision and goals, supported by stories and narratives of local history, reached immunity from criticism.

In examining these relationships we first trace the formation of the eTampere programme network in terms of a process of translation from the very beginning, to the mobilization of allies. In the limited space available, we have to focus our attention mainly on the first two stages of the translation: how the programme vision and goals came to justify and motivate the existence of the network, and how the core actors got involved in the programme network. The empirical materials we employ in the analysis consist of various written documents, including newspaper articles, programme plans, documents and evaluations, as well as interviews with key actors of the eTampere programme. In the spirit of actor-network theory, we have traced from the documents aspects that have performed the essential role of intermediary in the formation of the programme network. Moreover, we have selected the interviewees by following them through the network with the help of one interviewee leading us to another. All ten interviewees played key roles for instance as members of the programme planning team or as sub-programme leaders.

The interviews were thematically and chronologically structured, starting from where and how the interviewee had first heard about the programme. The interview did not follow exact questions, rather themes for a wider discussion about networking within the programme. This proved to be the right method as the interviewees were generally very communicative and enthusiastic about the topic. It turned out that the eTampere programme was still important to the interviewees and they were proud of having participated in the programme implementation.

The interviews focused particularly on the formation of the network including the processes of recruiting, interaction, communication and cooperation. The significance of defined goals, and the stories related to them, became evident directly or indirectly in every interview. In the following, these important themes are analysed through three tightly interrelated contexts. Firstly, we focus on the recruitment of new participants and the marketing of the programme idea. While the recruitment mostly took the form of finding new organizations as partners for the programme, the process was typically based on personal contacts and the organizations were represented by individual actors.

Secondly, we look into how a sense of belonging and solidarity was created and maintained during the programme implementation. Finally, we discuss the role of the story of eTampere in creating group identity and in (re)writing the identity and image of the city of Tampere.

The Formation of the eTampere Network

eTampere was a nationally as well as internationally well-known high-profile technology programme gathering local development elites under its umbrella. Its goals were highly ambitious and they were meant to be reached through efficient cooperation and networking. The programme drew inspiration from the past but looked to the future. Its aims were reflected in the formation of the network, a process that we set out to analyse using the concepts of story, intermediary, translation and black box in the context of the cognitive dimension of social capital.

The beginnings of the eTampere programme can be traced back to spring 2000. At that time, individually held ideas started to coalesce and crystallize, and a number of actors interested in the ideas of information society sensed that the atmosphere might be right for this kind of grand initiative. A key actor in this initial stage was Juha Kostiainen, the director of Business Development Centre of the City of Tampere. He provided legitimacy for the programme because of his formal position in the city government (Lähteenmäki-Smith and Steineke 2003). Moreover, his capacity to gain support for the idea – to translate the interests of other actors – was further enhanced by his double role as a politician and learned expert regarding the city's development policy.

Two other individuals who became interested early on were an experienced information technology expert Klaus Oesch and Reijo Paajanen, a talented engineer whose achievements included participation in the development of the pioneering smartphone, Nokia Communicator. Oesch and Paajanen paved the way for eTampere by charting the local business enterprises' possible interests toward a large-scale e-programme and thus creating tools for successful translations of new actors. Soon, a number of experts and business elite members were convinced that their interests aligned with those of the programme idea, and they moved on to present it to the city mayor, Jarmo Rantanen. Once his support was obtained, the programme network started to take shape progressively. The preparatory group appointed by the mayor quickly outlined the programme, and after its formalization the plan was ready to meet a wider audience. The plan was very ambitious with the aim of making Tampere nothing less than a leading city in the global information society.

> eTampere will make Tampere a key city within the global information society by strengthening the knowledge base, generating new businesses, and creating new public online services available to all citizens. (eTampere Programme Plan 2000)

The next step was to set up a planning team of nine persons, who also acted as a working committee for the programme with almost unchanged composition. The team was formed quickly and in addition to Kostiainen, Oesch and Paajanen it consisted of six information technology experts affiliated either with the Tampere University of Technology, the University of Tampere, or local research and development organizations. The core actor-network took shape without much friction not only because the participants found their interest in alignment with those of the programme, but also because they mostly knew each other. This illustrates well the 'Bourdieuian' understanding of the central role of social connections as assets that can be mobilized to achieve goals in complex and contested situations. In the end, the core actors were all willing to contribute to the programme and to the development of the city of Tampere. While some felt that the eTampere programme was perhaps too ambitious, aspirational and without a clear content, it nevertheless seemed like a one-time opportunity not to be missed.

> I think we felt that now we were on to something completely new, something big and significant, and the common spirit arose from there, that we are not just dealing with the old familiar things. We really had people willing to take giant steps forward, and that was the thing. We were at the Tampere Hall, where a wide range of interest groups were summoned together, a few hundred people. I gave the presentation speech, and we went through those subprogrammes. We had this aim to be a world leader in this and that, and I started to wonder whether I should ask if we *really* are into this. So I decided to ask. I was a little anxious but then I asked those who believe that we can reach these goals to raise their hands … rector Jorma Sipilä and others in the front row… a few hands rose … Finally everyone had their hands up, and it was evident that now all these businessmen, public administration and leaders, they all have their hands up believing in this. It was something stunning. We tend to have a somehow laconic and inflexible atmosphere where everyone has it that 'no, we can't do this' when one tries something. So we were rather surprised because we were not expecting this kind of eagerness. (Interview with Juha Kostiainen, director of Business Development Center of the City of Tampere)

Despite the spectacular launch of the programme in January 2001, broadening the eTampere network did not always run smooth and easy. Swift expansion was a vital goal because the key idea was that the programme would not merely finance new activities but also bring together existing ones. The challenge, then, was to persuade a broad array of actors with dissimilar backgrounds and interests to embrace the idea of added value emerging from their participation in the programme (Lähteenmäki-Smith and Steineke 2003). However, some consensus-building measures notwithstanding, the reception of the programme idea and its goals turned less unreserved and the core actors had difficulties in gaining a strong backing from their respective stakeholders. To work out properly a broad-based collaborative programme, such as eTampere, it would depend on actors who could credibly represent the business and the research communities as well as the public sector (Lähteenmäki-Smith and Steineke 2003). Now it turned out difficult for the members of the planning group to achieve successive translations that would extend the network outwards by mobilizing actors beyond the circle of 'believers'.

This failure can partly be accounted for by looking at the cognitive dimension of social capital and especially the factors that influence interaction in actor-networks. Firstly, the swift progression from the programme's public announcement to its implementation left little time for establishing trustful relationships and this was reflected in the initial suspicions towards the programme. Secondly, the eTampere programme was promoted as an 'umbrella' under which the interested parties could gather with their common interests. This image failed to highlight the aspect of local interdependence between the actors and organizations involved, and thus weakened the social basis of the network formation. It also pointed towards an open idea of networking that did not seem to direct social interaction or foster the development of dense collaborative relationships.

Put in terms of actor-network theory, the programme vision and its goals were at first unable to win over the more suspicious actors, and hence their translation remained incomplete. The relative shortage of time for establishing network relationships fueled the sense that the parties had different values, languages and ways of doing things. This hindered the possibility of mobilizing trust between the core actors and their stakeholders, as evidenced by the difficulty to find new partners willing to express full commitment to the programme, as well as by the divergent understandings concerning how realistic the programme goals were and how much time it would take to reach them.

> What was a good thing to learn in eTampere was that if you want people to co-operate with you, you have to do things for them. Our wider aim was to build an information society with our six subprogrammes including technological, financial and social dimensions, and this was too 'far-out' for average citizens in the Tammela market square, and for many other people as well. So our eTampere programme, and particularly the rhetoric and 'verbal play' peculiar to it, were quite challenging to transmit to different fields. (Interview with Jarmo Viteli, eTampere programme director)

In other words, the programme vision, as conveyed by the key actors, required that the potential new partners identify themselves in a very particular guise as part of the 'success story' of eTampere. Those who failed to connect with the image of a dynamic programme striving for a fantastic future found themselves in the periphery of the network and were inclined to opt out. A case in point is Jari Seppälä, the communications manager of the city of Tampere, who remained suspicious about the goal of Tampere becoming world leader in the ICT sector (Vuorikoski 2001). Criticism concerning the fanciful and elitist tone of eTampere was made public in autumn 2001 on the pages of the local major newspaper *Aamulehti* where a number of citizens expressed their disappointment with a programme that referred to the ordinary citizens as the foremost participants and winners in eTampere, but failed to engage actors outside the narrow elite.

Some frictions notwithstanding, by the end of 2002, the eTampere programme had established a governance organization comprising of a steering committee, a working group, and eTampere office located in the renovated Finlayson factory area. The network had extended to cover six subprogrammes including the *Information Society Institute* (ISI), *eBusiness Research Centre* (eBRC), *Research and Evaluation Laboratory* (RELab), *eAccelerator*, *Technology Engine Programmes* and *Infocity*. Moreover, the programme had successfully launched 11 projects with a total volume of €19 million, secured financing from the City of Tampere totalling almost €6 million, and prepared project funding applications with a total volume of €45 million (Lähteenmäki-Smith and Steineke 2003). Apparently, the programme was advancing relatively well (Figure 5.4). In accounting for possible reasons for this success, we argue that while a precarious resource, the 'power of stories' had much to do with it.

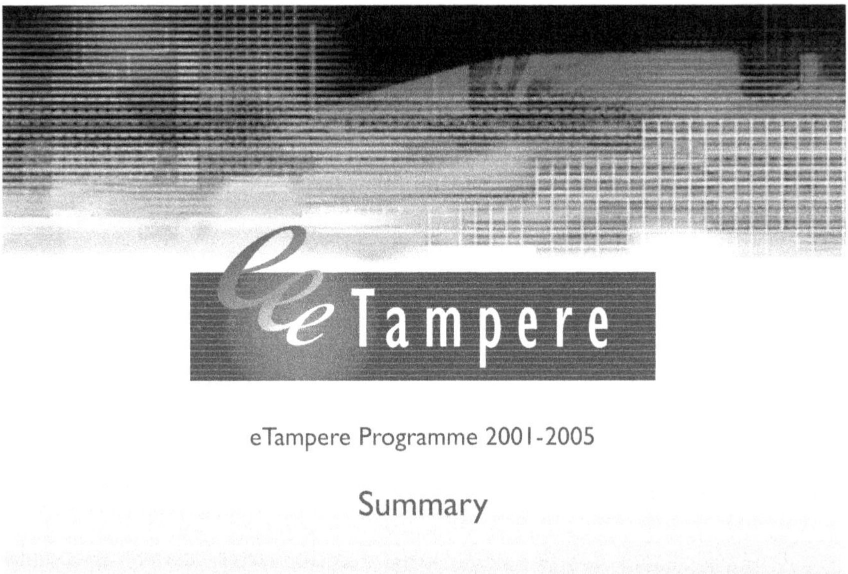

Figure 5.4 The logo of eTampere taken from the programme's final report

Source: eTampere programme summary 2005.

The story of eTampere, its vision and goals, figured in the dynamics of the programme network formation in two partly different ways. Firstly, the recruitment of new network members were mostly made in individual face-to-face conversations where it was possible for the network protagonists to effectively communicate their eagerness and confidence concerning the idea of eTampere, and thus win over the trust of new actors.

> The programme idea was so significant and powerful as a brand that I became encouraged to come along with a full contribution. (Interview with Hannu Eskola, eTampere subprogramme leader)

This highlights the significance of the programme vision told in the form of a story, a narrative that fosters a sense of membership and belonging. The alignment of new actors' interests with those of the programme network is much facilitated by elements and experiences that both parties share in the narrative of the programme as an extension of local history. We return to the significance of the link between individual and collective identity in the next section.

Secondly, more popular acceptance for the programme was acquired through a number of public seminars and briefings. These too, turned out to be successful as the enthusiasm of the motivated speakers converted more and more people to their side. Among the foremost spokespeople for eTampere was director Juha Kostiainen, who made a point of referring to key figures in the history of Tampere when presenting the programme to the public. One point of reference was the former rector of the University of Tampere, Paavo Koli, who was known for his bold vision and who now embodied all the dynamism and courage that was held to be characteristic of Tampere's identity.

> There is no longer a point to contending with poor alternatives, such as a happy existence as an average regional centre. The main reason for everyone's excitement was the feeling that we are really doing something remarkable and that we really are into it. This shows how powerful the spirit can be. It is not bureaucratic business as usual but believing and doing hard work. And – as a matter of fact – if the goals and the rhetoric would have been different, we would not have reached anything. (Interview with Juha Kostiainen, the director of Business Development Centre of the City of Tampere)

At this stage, any remaining signs of modesty had vanished. The stated goal was to lift the city to a world leading position in the ICT sector, to take the opportunity of showing the way and heading for the future boldly. This rhetoric demanded that some aspects of Tampere and her history are brought to the fore while others are eclipsed, i.e. that the identity of the city is established and portrayed in particular ways.

Black-Boxing Tampere for eTampere

In addition to the dynamics of network formation, it is useful to highlight means that are available for stabilizing the actor-network. Studies concerning the history of Tampere and the city's development policies have highlighted recurring aspects that can be seen as part of the city's development continuum, a path to which the eTampere programme also belongs (Haapala 2005; Kautonen 2006). Active strategic adaptation has been characteristic of Tampere as the city government has sought to prepare for the future by detecting socio-economic trends and changes and by adapting to those that have parallels with local resources and strengths. However, in addition to a level-headed outlook on future challenges, development efforts in Tampere have often evoked a sense of drama that stems from crises, great opportunities, or appealing visions, and mobilizes individuals and coalitions to pursue change (Kostiainen and Sotarauta 2003, 433).

It is here that the concept of the black box becomes useful (Latour 1987). By black-boxing assumptions concerning the network participants' individual or collective identity, the key actors were able to reduce the risk of harmful criticism against the social basis of the emerging network. As seen in the case of identity building for the eTampere programme, this not only made the network more durable but also enabled more efficient and concerted action. Looking at the process in terms of black boxing is important also for understanding how 'identity talk' functioned as a source of the network's power (Åkerman 2006, 40–41). To this end, it is worthwhile taking a look at the manner in which the programme idea was promoted both inside and outside the network, and thereby extending the analysis beyond the programme and its goals.

Reflecting the enthusiastic atmosphere among the elites involved in eTampere networking, a will to adjust the urban identity of Tampere surfaced in early 2000. While based on the 'obligating', time-honoured image of Tampere as the largest inland industrial centre in Finland and a progressively-minded city, four new elements were suggested as necessary changes to the traditional image. Firstly, Tampere should sustain her status as the *second most important city* in the country, after the Helsinki metropolitan region. This would ensure the city's domestic competitiveness as a site for investment and innovation, particularly in relation to the equally-sized city of Turku that at the time was pursuing competing ICT-related development policies (Ahlqvist 2008). Another important facet in this claim for (relative) centrality was the perceived need amongst the programme protagonists to ascertain that Tampere has what it takes to reach the visibility and recognition the city was striving for.

> Tampere is in an excellent position to become the key city within the global Information Society. Tampere has world-class research in the field of ICT and it has a tradition of high-quality research in the fields of social sciences and communications. A large number of well-known ICT and communications companies, such as Nokia,

> Sonera and Alma Media, are located in Tampere, as are their product development units. The city of Tampere has always been a leading city in Finland when it comes to the development of public online services. (eTampere Programme Plan 2000)

Secondly, the city has to improve her *competitiveness* in the face of globalization. This claim was rooted in the fashionable theories emphasizing cities' ability to attract various 'flows' and build networks as key elements in the global competition for ICT-related activities and investments. For Tampere, it would not suffice to maintain her prominent position in the domestic competition, but also gain a foothold among urban locales on the global scale.

> Professor Manuel Castells, one of the leading researchers in the Information Society, describes the new economic order in his book *The Rise of the Network Society* … by saying that we have moved from a space of places to a space of flows. … The dynamics of the various flows present totally new challenges to cities and regions. As the flows can alter their currents very quickly indeed, cities need to continually develop their competitiveness and attractive features. A city cannot attract all flows, which is why they must consider which flows they wish to attract. Competition for flows takes place at a global level. (eTampere Programme Plan 2000)

Thirdly, the city's *identity has to be updated* in conjunction with the first two processes. This is the moment where the collective identity of Tampere, negotiated over the course of several decades and centuries through myriad social practices and contested discourses, is yet again taken up as a vessel for political rhetoric and persuasion. The need to 'revise' a city's identity frequently comprises of attempts to reshuffle and reorder not only meanings and attributes of the city's representations, but also the subjectivities, power relations and political interests interrelated with these representations.

> A global Information Society means the following for cities: … Occupations are continuously disappearing and changing. Competition between companies, regions and countries becomes fiercer. Regions will compete over the location of companies and people with special knowledge. Subsidized politics is replaced by the politics of competition as the role of created production factors increases. … Infocity [subprogramme] will spread the eTampere programme to the everyday life of Tampere citizens. … The whole city of Tampere will become one big testing and development environment for domestic and international players. (eTampere Programme Plan 2000)

Fourthly, this new identity has to be based on success in fostering the *development of the ICT sector* in Tampere. This assertion draws attention to the 'necessary' and 'inevitable' direction that the city's development should follow if she wishes to sustain her place on the map of notable and prospering urban locales both nationally and internationally. In regard to this emphasis, it seems no accident that the eTampere programme office was located in the renovated Finlayson factory area hosting dozens of ICT and media businesses (Figure 5.5).

> It is generally believed that developed economies are moving from an industrial to a post-industrial society. The new social paradigm is called the Information Society. One of the features of the Information Society is the increasing importance of information as a factor of production. … The Tampere Urban Region has a clear strategic view of its future. It wishes to become a centre for training, research, culture and knowledge-based business. Mechanical engineering and automation, information and communication technology (ICT), health care technology and media services have been chosen as special areas of focus. (eTampere Programme Plan 2000)

Figure 5.5 The eTampere office was located within the new ICT landscape of the Finlayson area (photo by Jouni Häkli)

These four new factors, together with the traditional core identity, were 'black boxed' in the course of the programme life cycle so that eventually it was very difficult to call it into question even in the more critical appraisals of the eTampere programme. Consequently, as the information technology sector was going through its most successful years with huge expectations and visions for the future, it seemed that this was the right time for the city of Tampere to launch something new and ambitious.

> On the other hand, being the 'second city' is stressed here in Tampere. It is always the capital versus the rest of Finland, and for the 'second city' it is very important to be appreciated. Therefore we must succeed, and for that you must have direct connections to the outside world: you can't do all business through Helsinki. In Tampere we have Ryanair, and some Blue1 flights too. I think we have a kind of readiness to do larger things than we are supposed to. We want to be seen and heard, not just a typical Finnish city, and we must aim higher. That brings us back to eTampere, which is a good example of brave new ideas, built on a strong basis of cooperation. (Interview with Jarmo Viteli, eTampere Programme Director)

> It is typical of Tampere, that the Pirkkahalli conference center must be bigger than the one in Lahti, the Tampere Hall must have more capacity compared to a similar building in Helsinki, Näsineula tower must be a little higher than the tower in Kuopio etc. We are used to do things; we want to be the number one. We have this 'American dream' in a way that we must follow through. We are on the cutting edge of information society development, the number one in the whole world. We have world market leader companies in mechanical engineering and many world scale innovations have been launched first here in Tampere. This is typical for Tampere, and we want to keep it up and add some momentum to information society development. We do not want to concentrate only on research, innovations and bringing new products to market. We want to get closer to the people. (Interview with Olli Niemi, Manager of Tampere Technology Centre Hermia)

This 'identity talk' obtained enough support, and in the official context of the eTampere programme it developed a life of its own through the programme's explicit vision and goals. Established early on, they created a specific image for both eTampere and the city of Tampere, which eventually developed into a kind of brand. While efforts were made to make the ideas and aims of the programme more concrete, visible and understandable for the citizens, the new identity of Tampere was assiduously discussed and repeated both in public and amongst the programme actors to the degree that it turned into a mantra. This black-boxing of the identity of Tampere helped the emerging actor network in arguing for and implementing the programme, and thus gave the network more durability and stability during the programme lifespan. It also assisted in the crucial stages of translation through which the programme network eventually took shape.

Conclusions: Translating the eTampere Network

Open-mindedness, extensive cross-disciplinary and cross-organizational cooperation and networking are widely used keywords in the discussion of development projects and programmes and their factors of success (Hadjimichalis and Hudson 2006). Striving for such ideals and adopting new ways of thinking and acting demand a lot from actors as well as from their potential collaborators. Shared visions may first look attractive and ambitious and implementation plans may seem to be well thought out, but often the theory fails the test of practice. Heterogeneous actors with their different values, thinking schemes and ways of doing things may eventually conflict, especially after the initial stage of network formation. However, networks that succeed in finding, engaging, enrolling and mobilizing new partners, remain the cornerstones of successful cooperation and are likely to leave these difficulties behind (see Cooke et al. 2005). This capacity is to a great extent, dependent on cognitive social capital, together with strategic success in creating stable and durable actor networks capable of contributing to the field of development in question.

In this chapter we have examined the role of common and shared visions and goals in building and maintaining social capital amongst participants in the ambitious eTampere development programme. We have focused more on the social prerequisites of trust and social capital than their enactment in the processes of eTampere networking. However, by way of concluding we also wish to extend our discussion towards the latter aspect. To this end, we highlight some key moments where social capital and trust (or the lack thereof) have critically influenced the successes and failures of the eTampere programme.

Firstly, when the eTampere network strove to expand across organizational and disciplinary boundaries, social capital as a context-dependent asset became a critically important but scarce resource. This had much to do with the issues of time and interaction – two essential factors shaping the evolution of social relationships. In the case of eTampere, the dense social capital and trust amongst the core participants, combined with the short interaction history between new members on the fringes of the programme network, made it difficult to achieve a strong basis for network building throughout. Consequently, networking faced difficulties where the core members could not enact their social capital in enrolling new participants, but instead had to resort to the programme vision in establishing confidence toward eTampere.

This problem was accentuated on the fringes of the eTampere network, particularly in attempts to establish cooperation with businesses and firms. According to the programme evaluation, cooperation with business life never reached the intended level and the reasons can be traced back to the fact that private businesses typically seek direct advantage from cooperation, and they operate through concrete short-term goals (Niinikoski et al. 2006, 43). This business way of doing things lies in stark contrast with the eTampere programme vision that was expressed in deliberately broad terms to sustain its appeal. Consequently, it was difficult for the aspiring new members of the programme network to gain a concrete understanding of how the programme vision could be implemented and realized, and how the highly ambitious goals might be reached. Arguably, the main vision remained too abstract and devoid of clearly formulated and concrete short-term goals through which the actors could have associated with the larger aims of the programme (Manninen 2005; Niinikoski et al. 2006).

> More active and committed participation by firms and companies has not been achieved (Seppo Haataja Director, Nokia Enterprise Solutions; cited in Manninen 2005).

> In the beginning, the goal was too vague and later on all achievements in the field [of the ICT sector development] have been credited to eTampere. (Kimmo Kauranen Region Manager, Fujitsu Services; cited in Manninen 2005)

Secondly, building the eTampere network involved a set of tightly interconnected translations that followed the stages analytically distinguished by Callon (1986). Ultimately, networking under the eTampere programme was successful with translations. The story behind the programme vision was well-planned and formulated, and the programme attracted a great deal of attention, especially at local and regional levels. At an initial stage, key actors gathered around the city mayor Jarmo Rantanen, whose authorized position acted as an intermediary in translating the actors into an embryonic network (Lähteenmäki-Smith and Steineke 2003). The situation of Tampere was *problematized* in the face of globalization, which provided the manuscript for further translations by defining the identities and interests of other actors in relation to the problem. With success in *interessement*, this core network was then established as the planning group for the programme. The emerging story of Tampere as a world leader in ICT development acted as a key intermediary, alongside the compelling physical setting provided by the city landmark, the Tampere Hall conference center. The established planning group network drafted the eTampere programme, which thus became evident and tangible as a highly authorized document that further acted as a powerful intermediary with its persuasive story and vision of the future Tampere. This document, supported by the official programme website, was an important intermediary in all further translations through which new participants and partners were *enrolled* into the eTampere cooperation. Success in securing that new network participants accept the roles that have been defined for them, and then *mobilizing* these actors, completed the initial set of translations.

However, some translations were left incomplete and the programme did not expand quite as effortlessly as had been expected.

> We did not prepare the programme well enough. We should have engaged more actors in close cooperation already at the planning stage. On the other hand, business life did not make many concrete proposals for plan of actions (Jarmo Viteli, eTampere programme director; cited in Manninen 2005). Companies were under the impression that they can not participate. In compiling a programme of such merit, we should have better utilized businesses that were ready to cooperate. (Eero Ottila, Region Manager in Fujitsu Siemens; cited in Manninen 2005)

Reasons for the failure of some translations can be partly accounted for by a lack of social capital at crucial stages of the network building. While the cognitive dimension of social capital worked well to stabilize and align the core members of the programme planning group, problems of translation occurred along with network expansion when enrolment outside the like-minded actors began. At that stage, differences in working cultures – in ways of thinking and acting – accentuated to the degree that it was increasingly difficult to mobilize new network members to act towards the goals of the programme. Consequently, the idea of wide, multifaceted, cross-disciplinary and public-private partnerships, tied together by strong and weak ties, did not work in practice as had been intended. A consensus on participation was only partially reached, so that besides the core actors and their nearest collaborators, the programme network turned unstable and porous. In all, it was evident that the initial excitement failed to reach the fringes of the eTampere programme network where there was not enough dedication or a shared basis for cooperation, i.e. common and shared language, values and interests, or the time to create the cognitive basis of social capital.

Trust-related processes are affected by similar factors that shape social relationships, and as we have seen in the case of eTampere, the processes of trust formation also face significant challenges in temporary, project-oriented networks. This presents a real challenge for the emergence of social capital as a resource that the networking actors can enact as an asset. From the viewpoint of engagement, fixed term projects and programmes are problematic because network participants usually have to start establishing their after-programme life while the project is still ongoing. Particularly when no concrete signs of the project succeeding or reaching its short term goals are evident, the initial enthusiasm tends to wane. This was the case also in eTampere with projects that only had long-term goals, which were not defined concretely and were not even expected to be reached during the programme life span (eTampere Programme Summary 2005).

Where the programme turned out to be more successful was the manner in which its key actors were able to mobilize a particular vision rooted in the history of Tampere as a progressive-minded industrial centre. It was largely the power of the story of eTampere as a 'natural' extension of the city's development history that helped overcome skepticism that many actors initially held concerning the programme's magnitude and grandiose goals. The story, as inscribed in numerous documents, reports and presentations concerning the eTampere programme, acted as an intermediary in the network formation, fostering the successive stages of translations that were necessary for the programme to reach viable dimensions. This helps in understanding how the eTampere programme, while failing to boost the city into a world leader position, nevertheless managed to reach the goal of amassing and mobilizing as many as 400 ICT-related projects with a total volume exceeding 130 million euro (Niinikoski et al. 2006).

As an 'intermediary', the story of eTampere functioned like a frame of thought through which actors came to observe and interpret their environment and relate it to the hoped-for future of the city. It played a key role in the processes of making choices and decisions concerning participation in the programme network. It also had effects on the formation of shared meanings, interpretations, and ways of conducting business, thus resembling a 'local epistemic community' – a relatively homogeneous social space comprising of a shared understanding of the world, compatible backgrounds, training and identity (Sheppard 2005, 153). This turned out to be a significant force because it eventually enabled concerted action by encouraging the development of trusting relationships between heterogeneous actors under the eTampere umbrella.

The (re)production of the story of eTampere also functioned as a significant source of social power, not least because 'members of any epistemic community typically find it hard to see the planks in their own eye; indeed, this may be a prerequisite for membership' (Sheppard 2005, 156). Like in all actor-networks, in eTampere, social capital stood not as a benevolent collective resource that is equally available for all, but an unevenly distributed set of actual or potential resources embedded within and available through the social relationships and contacts that actors can enact. While power is dispersed in such networks, it can be directed by some social actors more than others (Castree 2002, 135). As a resource, social capital can be harnessed to set off policy processes that require large scale public funding as was the case with the eTampere programme. Three years after the formal ending of this ambitious programme, it still remains to be seen what part of its networked 'capital' may truly become 'social' – how the great majority of Tampere denizens may come to benefit from the achievements of the few.

Bibliography

Ahlqvist, T. (2008), *Technologies of regional transformation: The construction of an information economy in the region of Southwest Finland 1985–2001 (Summary)*, Annales Universitatis Turkuensis, series C 266 (Turku: University of Turku).

Åkerman, M. (2006), *Tiedon tuotannon käytännöt ja ympäristöpoliittinen toimijuus* [Practices of knowledge production and environmental agency],

Acta Universitatis Tamperensis 1139 (Tampere: Tampere University Press).
Bijker, W.E. and Law, J. (eds) (1992), *Shaping Technology/Building Society: Studies in Sociotechnical Change* (Cambridge, MA: The MIT Press).
Boje, D. (1999), 'Hegemonic Stories and Encounters between Storytelling Organizations', *Journal of Management Inquiry* 8:4, 340–60.
Bourdieu, P. (1986), 'The forms of capital', in J.G. Richardson (ed.).
Burt, R. (1992), *Structural Holes: The Social Structure of Competition* (Cambridge, MA: Harvard University Press).
Callon, M. (1986), 'Some elements of the sociology of translations: domestication of the scallops and fishermen of St Brieuc Bay', in J. Law (ed.).
Callon, M. (1991), 'Techno-economic networks and irreversibility', in J. Law (ed.).
Callon, M. (2001), 'Actor Network Theory', in N. Smelser and P. Baltes (eds).
Camagni, R. (ed.) (1991), *Innovation Networks: Spatial Perspectives* (London: Belhaven Press).
Castells, M. (1996), *The Rise of the Network Society – The Information Age: Economy, Society and Culture* (Malden, MA: Blackwell).
Castree, N. (2002), 'False antitheses? Marxism, nature and actor networks', *Antipode* 34:1, 111–46.
Cohen, D. and Prusak, L. (2001), *In Good Company. How Social Capital Makes The Organizations Work* (Boston, MA: Harvard Business School Press).
Coleman, J. (1990), *Foundations of Social Theory* (Cambridge, MA: Harvard University Press).
Cooke, P., Clifton N. and Oleaga M. (2005), 'Social capital, firm embeddedness and regional development', *Regional Studies* 39:8, 1065–77.
Czarniawaska, B. (1997), *Narrating the Organization Dramas of Institutional Identity* (Chicago and Longon: University of Chicago Press).
Czarniawaska, B. (1998, *A Narrative Approach to Organization Studies*, Qualitative Research Methods Series 43 (Thousand Oaks, CA: Sage Publications).
eTampere Programme Plan (2000), <http://www.etampere.fi/etre_old/projektit/pics/131.pdf>, accessed 20 November 2007.
eTampere programme summary (2005), <http://www.etampere.fi>, accessed 20 November 2007.
eTampere website (2007), <http://www.etampere.fi/historia/etampere_historia>, accessed 20 November, 2007.
Florida, R. (2002), *The Rise of Creative Class and How it is Transforming Work, Leisure, Community and Everyday Life* (New York: Basic Books).
Gardner, H. (1995), *Leading Minds. An Anatomy of Leadership* (New York: Basic Books).
Granovetter, M. (1992), 'Problems of explanation in economic sociology', in N. Nohria and R. Eccles (eds).

Fine, B. (1999), 'The developmental state is dead: long live social capital?', *Development and Change* 30:1, 1-19.
Haapala, P. (2005), 'History of Tampere: The very long road to Informational City', in A. Kasvio and A.-V. Anttiroiko (eds).
Hacking, I. (1999), *The Social Construction of What?* (Cambridge, MA: Harvard University Press).
Hadjimichalis, C. and Hudson, R. (2006), 'Networks, regional development and democratic control', *International Journal of Urban and Regional Research* 30:4, 858–72.
Inkpen, A.C. and Tsang, E.W.K. (2005), 'Social capital, networks and knowledge Transfer', *Academy of Management Review* 30:1, 146–65.
Kasvio, A. and Anttiroiko, A.-V. (eds) (2005), *e-City – Analysing Efforts to Generate Local Dynamism in the City of Tampere* (Tampere: Tampere University Press).
Kautonen, M. (2006), *The Regional Innovation System Bottom-up: A Finnish Perspective*, Acta Universitatis Tamperensis 1167 (Tampere: Tampere University Press).
Kostiainen, J. and Sotarauta, M. (2003), 'Great leap or long march to knowledge economy: institutions, actors and resources in the development of Tampere, Finland', *European Planning Studies* 11:4, 415–38.
Latour, B. (1987), *Science in Action. How to Follow Scientists and Engineers Through Society* (Milton Keynes: Open University Press).
Latour, B. (1993), *We Have Never Been Modern* (New York: Harvester Wheatsheaf).
Latour, B. (1999), 'On recalling ANT', in J. Law and J. Hassard (eds).
Law, J. (ed.) (1986), *Power, Action and Belief. A New Sociology of Knowledge?* (London: Routledge).
Law, J. (ed.) (1991), *A Sociology of Monsters* (London: Routledge).
Law, J. and Hassard, J. (eds) (1999), *Actor Network Theory and After* (Oxford: Blackwell).
Lehtimäki, H. (2000), *Strategiatarina kaupungista ja sen toimijoista* [Strategy narrative of a city and its actors], Acta Universitatis Tamperensis 746 (Tampere: Tampere University Press).
Lähteenmäki-Smith, K. and Steineke, J. M. (2003), 'eTampere: Building an electronic pathway to the future? Evaluation of eTampere' (Nordregio EP 2003:4), <http://www.nordregio.se/publications.htm#2003>, accessed 20 October 2007.
Manninen, J. (2005), 'eTampere: 132 miljoonan kiistakapula' [eTampere: a 132 million euro bone of contention], *Aamulehti*, 15 May.
Mishler, E. (1999), *Storylines: Craftartists' Narratives of Identity* (Cambridge, MA: Harvard University Press).
Murdoch, J. (1997), 'Towards a geography of heterogeneous associations', *Progress in Human Geography* 21:3, 321–37.
Murdoch, J. (1998), 'Of Actor Network Theory', *Geoforum* 29:4, 357–74.

Myllyntaus, T. (1991), 'The transfer of electrical technology to Finland, 1870–1930', *Technology and Culture* 32:2, 293–317.

Nahapiet, J. and Ghoshal, S. (1998), 'Social capital, intellectual capital and organizational advantage', *Academy of Management Review* 23:2, 242–66.

Niinikoski, M.-L., Uusikylä, P., Wiik, M. and Hjelt, M. (2006), 'eTampere-tietoyhteiskuntaohjelman loppuarviointi' [Evaluation of the eTampere information society programme], <www.etampere.fi/materiaalit/loppuraportti.pdf>, accessed 13 January 2008.

Nohria, N. and Eccles, R. (eds) (1992), *Networks and Organizations: Structure, Form and Action* (Boston MA: Harvard Business School Press).

Palmroth, A. (2004), *Käännösten kautta kollektiiviin. Tuuliosuuskunnat toimijaverkkoina* [To collectives through translations. Windmill cooperatives as actor-networks] (Jyväskylä: Jyväskylä University Printing House).

Pretzer, W.S. (ed). (2002), *Working at Inventing: Thomas A. Edison and the Menlo Park Experience* (Baltimore, MD: Johns Hopkins University Press).

Putnam, R. (1993), *Making Democracy Work: Civic Traditions in Modern Italy* (Princeton, NJ: Princeton University Press).

Putnam, R. (2000), *Bowling Alone: The Collapse and Revival of American Community* (New York: Simon & Schuster).

Richardson, J.G. (ed.) (1986), *Handbook of Theory and Research for the Sociology of Education* (New York: Greenwood).

Sheppard, E. (2005), 'Free trade: the very idea! From Manchester boosterism to global management', *Transactions of the Institute of British Geographers* 30:2, 151–72.

Smelser, N. and Baltes, P. (eds) (2001), *International Encyclopedia of Social and Behavioral Sciences* (Oxford: Elsevier).

Tsai, W. and Ghoshal, S. (1998), 'Social capital and value creation: the role of intrafirm networks', *Academy of Management Review* 41:4, 464–76.

Vuorikoski, V. (2001), 'Vain paras kelpaa e-Tampereelle' [Only the best is good enough for eTampere], *Aamulehti,* 15 January, 13.

Whatmore, S. (2006), 'Materialist returns: practising cultural geography in and for a more-than-human world', *Cultural Geographies* 13:4, 600–609.

Chapter 6
Networks and Trust in Venice: The Port as Social Agent

Stefano Soriani

Introduction

Over the last 20 years, the debate over Venice's future has been dominated by considerations of how to promote economic and social development in the area. The so-called 'Venice problem' is characterized by three aspects: the industrial decline of the Venetian lagoon; a set of environmental issues brought about by earlier forms of development; and the supposedly pernicious *commodification* of the city through processes of tourism. To date, the overwhelming aim of the literature tackling the Venice problem has been to foster development practices that take both the economic and environmental spheres into consideration – practices that don't just focus on cleaning up the polluted waters of the lagoon and on the environmental restoration of the area.

In relation to this, the port has played a crucial role. Ever since port activities were relocated to the mainland, they have had a profound impact on the development of the Venice area. The creation of the new port and industrial complex of Porto Marghera during the 1920s and 1930s, which involved reclaiming sections of the lagoon, propelled Venice into its great industrial modernity. In the process, the city was recontextualized by the lagoon, while the surrounding mainland areas were transformed by the capitalist dynamics that were restructuring the Italian economy at that time (Ernesti 2001; Soriani 1996). For many decades the port, which became one of the most important industrial developments in a coastal zone in Italy, had been the driving force behind the economic and social development of the Venice area. In the 1980s and 1990s however, the port and its associated industrial complex underwent a period of crisis, owing both to a decline in the formerly important coastal industries, and to the port's own inability to respond successfully to technological and organizational changes, which had revolutionized the transport sector and the very nature of port activity. The situation however changed again in the second half of the 1990s. While the productivity of the industrial component of the port continued to steadily decline, the commercial component began to show signs of a renewed dynamism. As a consequence, the port of Venice has once again become a vital force in the material and social transformation of Porto Marghera (Soriani and Zanetto 2002).

When considering this recent rejuvenation of the Venetian port, the role played by certain external factors must be acknowledged. During the time period in question, both the processes of economic globalization, as characterized by the growth in trade between Asia and Europe, and the associated restructuring of the global maritime industry significantly affected all Mediterranean seaports, as they opened the way to new opportunities and potential development. For the port of Venice and other major Italian ports in particular, another significant factor came in the form of new legislation that favoured the privatization of port management, as this provided the necessary conditions for improving their efficiency and thus competitiveness in the global market. It would be a mistake, however, to interpret the recent success of the port of Venice as resulting just from the opportune exploitation of shifts in global dynamics. Instead, as this chapter tries to show, this recent success stems from a complex process of re-evaluation and recontextualization of the local milieu, that took place in response to the renewed potential for commercial opportunities (see Governa 1997; Giaccaria in this volume). This process was made concrete through two connected means. Firstly, a fundamental transformation of the port system was mediated materially through the infrastructure and layout of the port. Secondly, the networks that defined the port sector as a social agent underwent transformation, re-establishing itself as a key economic and political player within the local milieu.

This chapter examines the different ways in which this complex process of transformation took place and has since been consolidated, at the same time as paying attention to the role played by 'narratives' and 'discourses' in promulgating the importance of the port for the future of Venice's development. In considering these issues, I also examine the relationship between the ways in which cooperative social networks were set up and consolidated and the local capacity to mobilize resources. This chapter attempts to answer the following questions: in what ways has the port of Venice changed as a social agent? What role has cooperation played in the mobilization of social capital and therefore also in the physical and social transformations of the port? What, exactly, are the factors that have facilitated the production and consolidation of these new cooperative social networks? And, finally, what are the broader geographies of these different relations?

Before moving on, however, two points need to be clarified. First, when Venice is referred to in this chapter, I mean the Municipality of Venice. It should be noted that as well as the historic Venice in the lagoon (also referred to as the 'historical city' or the lagoon city), the municipality of Venice is also comprised of Mestre, Marghera and Porto Marghera (the Venetian port and industrial complex), which are situated on the mainland (see Figure 6.1). This particular focus is justified by the fact that through the creation of Porto Marghera, the heart of the port of Venice was relocated to the Venetian mainland. Moreover, it was the creation of this new port on the mainland that led to the continual enlargement of the administrative boundaries of the city of Venice, such that it then encompassed mainland towns that had formerly been autonomous. The underlying logic here is that 'Venice's dominion stretches as far as its port extends' (Ernesti 2001; Soriani 1996).

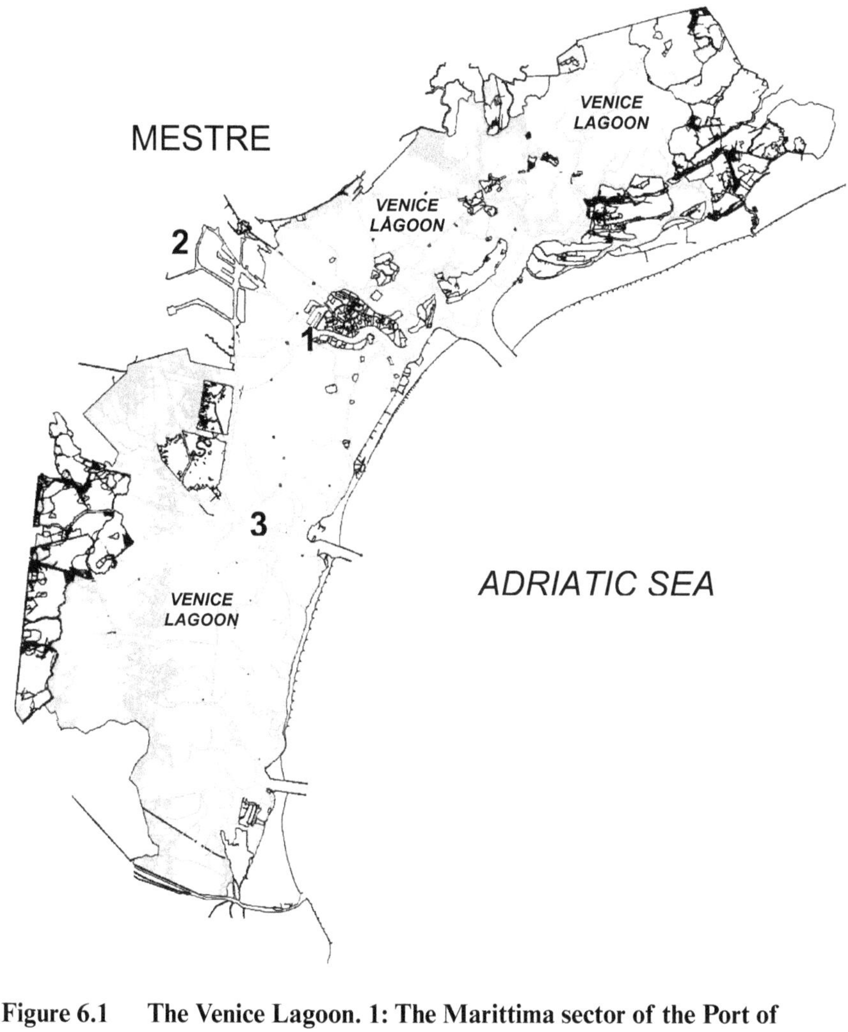

Figure 6.1 The Venice Lagoon. 1: The Marittima sector of the Port of Venice. 2: the Porto Marghera sector. 3: the Porto Marghera-Malamocco canal (the 'oil canal') (© Ministry of Infrastructures and Transports – Magistrato alle Acque through its Concessionary Consorzio Venezia Nuova)

The second point relates to the theoretical perspective taken in relation to the (re)production of networks and social capital in the context of Venice. It is important to note here that the belief that the industrial development of port of Venice is incompatible with the extremely fragile ecosystem of the lagoon has increasingly gained currency since the late 1960s. Additionally, the idea that Venice's future is dependent on the port and its associated industrial specialization has been effectively undermined by the impressive growth in the tourism industry and the ongoing decline of Porto Marghera's large industries. Unsurprisingly then, the issue of social legitimacy became increasingly important to the port in his role of local social agent. Against this background, from a social and political perspective, Venice during the 1970s and 1980s could be regarded as an 'arena' in which different views and visions of the city's future competed with one another, as did different social networks. Taking this into consideration, I adopt here Pierre Bourdieu's approach to social capital (Bourdieu 1986; 1998), since it regards the design and consolidation of networks as a 'tool' for attaining particular objectives. From this point of view, social networks can be interpreted as the outcome of strategies of social investment aimed at mobilizing resources and achieving goals. Accordingly, social capital loses its almost idyllic connotation of 'hidden treasure' – a connotation typically used in studies where social capital is linked to the concept of neighbourhood communities, or that contained in Putnam's (1993) reflections on the 'civicness' as the outcome of local history. Instead, it figures as a new type of productive resource for the realization of social, political and economic ends.

From a methodological point of view, this chapter critically examines technical reports, earlier studies and research projects, comments made both in the local and national press, and 'texts' that give voice to the principal actors involved. Attending at first to discussions surrounding the 'Venice problem' and the future of the city, I outline the economic and social context within which the port is currently operating and consider how the port itself has contributed to shape this context. I then go on to examine the factors that in recent years have helped to yield fresh opportunities for the port of Venice (such as globalization, geo-economic changes, and the privatization of port organization). Subsequently, I analyse the nature and mode of territorialization of these opportunities, in order to show how they were territorialized thanks to the formation and consolidation of a new Port Business Community (PBC), which was defined by far more complex social networks than those existed previously. In addition, I consider the role played by cooperation in the formation of these networks, which have a greater capacity to identify, interpret and put into practice new resources. The chapter then goes on to analyse the part played by 'narratives' about the port in the promotion and consolidation of the changes already noted. Finally, I offer some concluding thoughts on two particular factors: the role that the material and social changes (as promoted and consolidated by the PBC) have played, and continue to play, in the local debate concerning the future of Venice; and the contribution that an analysis of the Venetian case can make in relation to understandings of territorial dimensions as regards the mobilization of social capital.

'Past', 'Present' and 'Future' Venice

> Venice will end up being crushed by swarms of 'eat and run' tourists, long before it is flooded by high water. (Marco Michielli, President of the Veneto Hotels Federation, *La Nuova Venezia*, 10 November 2006)

> Venice is perfectly placed within what is considered a globalized society ... Its potentialities (i.e., its name and image, its huge stock of infrastructures, its large companies and cultural institutions, its production know-how) can all compete with the European capital cities, struggling for the location of European and world institutions, new businesses in the field of high-value and innovative industrial sectors, culture and communication. (D'Agostino 2002, 19)

As already noted, the social and economic problems connected with the process of tourist 'commodification' and the structural crisis of Porto Marghera's heavy industry, constitute two essential elements in the discussion concerning the future social and economic development of Venice.

With respect to the problem of tourist commodification, in 2006 approximately 3.5 million tourist stayed overnight in the lagoon city. Overnight tourists, however, only represent 30–35 per cent of the total number of visitors, which was estimated at more than 18 million in 2006. Clearly, day-trippers constitute the most important component of the phenomenon. This type of tourism impacts the city negatively in a number of different ways: it creates congestion and overcrowding in the most famous areas of the lagoon city, depletes its historical and architectural resources, pushes up the cost of living, and stimulates the city's depopulation.[1] Furthermore, this type of tourism does not encourage the establishment of new economic activities. Indeed, it actually facilitates the trivialization of the tourist sector itself, which is increasingly characterized by the 'eat and run' experience (Costa 1993; Musu 1998). Following this trend, the lagoon city is well on its way to becoming a tourist district *in toto*, and a badly managed one at that. In these circumstances, it is no surprise that the international press frequently suggests that the future of the lagoon city lies in its potential management as a *theme park*.

In terms of the decline in heavy industry, it should be remembered that, until the middle of the 1960s, Porto Marghera figured as one of the largest industrial areas in Mediterranean Europe and the only major concentration

1 There are 60,000 inhabitants in the historic center of Venice at present (with about 268,000 residents in the Municipality as a whole,) which is less than half the number of residents in the 1960s. A large number of these residents are in the older age-bracket.

of basic industries in the Veneto region (Soriani 1996). From the 1920s to the 1960s the port, whose basic function was the industrial transformation of imported raw materials, could be regarded as a single port and industrial complex. Acting as a vital force in the development of the Venice area, the port, in social terms, figured as a highly dynamic and powerful agent in the promotion and consolidation of social, political and economic networks on a local, regional and national scale (Ernesti 2001). Moreover, the boom phase that Northern Italy's economy witnessed during the 1950s and early 1960s supported the new design and further expansion of the port and industrial area (Soriani and Zanetto 2002). As is well known, however, an extremely high tide recorded on 4 November 1966 radically disrupted these plans. This in turn prompted a period of intense study and environmental campaigning, which placed more and more attention on the environmental implications of modernizing the port activities in the lagoon (linearization of port channels, realization of the 'oil canal', etc.).

With this increased environmental consciousness, the social climate changed radically, leading to a phase of deep ideological conflict, which saw those who supported modernization and industrial specialization of the port pitted against those who wanted to safeguard Venice and the lagoon environment from further economic development (Soriani and Zanetto 2002). Moreover, the work of international committees and the increasing media attention of events in Venice (such as industrial accidents, flooding and depopulation) reinforced the perception of Venice as a city that was destined to die. Indeed, these discourses helped to construct Venice as an icon for all the problems posed by technology and modern industrial development – an idea that gained currency particularly in the tourism sector. These factors set in motion a protracted process of re-interpretation and recontextualization of port development in the lagoon. The plans for further industrial and port expansion were first halted and eventually cancelled, with significant social consequences, especially for Porto Marghera, which was continually accused of being the source of Venice's major environmental problems (Soriani 2004).

If the critical interpretation of the Fordist experience in the Lagoon began with an increasing awareness of environmental problems in general and with the emergence of a global debate about 'safeguarding' the lagoon city, the crisis and difficult restructuring of heavy industry during the 1980s deepened this anti-expansionist sentiment (Zanetto and Tassan 1990). After all, the port's commercial activities were plagued by escalating problems at this time, due to local organizational inefficiency and to the increasingly peripheral status of the Northern Adriatic Sea, which resulted from the reorganization of maritime and continental shipping routes during the emerging 'intermodal era' in the transport and distribution industry. In fact, the port of Venice played only a small part in the difficult containerization phase in the Italian ports. Having entered the container shipping market in 1975, by 1983 the port handled 106,000 metric tons of freight, an amount that increased in 1993 to 117,000 metric tons

(Soriani and Zanetto 2002). Further proof of the Venetian port crisis during that time is apparent in the fact that no significant renovation of the port's infrastructure was carried out during the 1980s and early 1990s. In addition, the failure to maintain and dredge the port channels led to an increasing reduction in marine accessibility. Essentially, this was a period of deep distrust within the port sector as it became socially and politically marginalized, with Porto Marghera increasingly being perceived as a difficult legacy inherited from the past – a kind of 'accident of history' (Musu 1998; Soriani 2004). The spiraling crisis in the industrial sector continues today as the major problems of soil and water pollution, combined with the complex financial and legal framework associated with its clean up issue, have seriously restricted the process of economic revitalization (Mannino et al. 2004).

Recent debates about the future of Venice have typically focused on, or have been sustained by, 'narratives' and visions of its possible role in the changing global economy and society (Costa 1988; D'Agostino 2004; De Rita 1993; Indovina 2003; Musu 1998). The principal reasoning behind these discussions is that Venice has potential resources which could, at the sunset of the Fordist era, sustain a new and dynamic role for the city on the world stage. The following intertwined issues have been foregrounded in debates concerning the future development of Venice: a) the tourist industry and its activities must be revitalized and better organized, given that tourism has become, in the words of former Major Paolo Costa, the 'real Venice' (Costa 1993); b) the formation and consolidation of a new entrepreneurial elite must be encouraged; c) the resources and image of the lagoon city (its culture, history and architecture) can be understood as consistent with the aim of developing new 'immaterial' activities on an international level of excellence; and d) commercial, port and maritime-related activities (physical distribution, services, logistics, etc.) in conjunction with one another could offer essential levers for revitalizing Porto Marghera in accordance with the driving forces of the global economy. Additionally, a great deal of attention has been paid both to promoting new forms of metropolitan government in the heart of the Veneto region (as a means of rescaling and re-mapping the problems and opportunities in Venice, and as a lever for improving the competitiveness of the Veneto on the European scale), and to the development of new forms of strategic planning.

For the purposes of this chapter it is important to underline two particular themes for discussion. First, the debate about the 'future' Venice has increasingly focused on the relevance of promoting new cooperative attitudes and approaches between actors (individual and collective, private and public). Political authorities and the most influential urban elite quite frequently urge all local public and private institutional bodies to find more effective modes of cooperation. The assumption here is that, thanks to the economic and social complexity of the Venice situation, only greater cooperation between social agents and networks can mobilize new resources and thus 'make things work'. The second theme concerns the possible role of the port in the future development of Venice.

From the current discussions on this subject, and from the principal reports and documents that sustain the local debate about the future of the city, it emerges that after the period of distrust and the profound crisis concerning its social legitimacy, the port is once more perceived and portrayed as a dynamic force for social and economic revitalization. In particular, commercial port activities and port-related services are regarded as fundamental driving forces in the attempt to counterbalance the process of tourist commodification.

The Potential Emerging Opportunities for the Port of Venice: Globalization, Geo-economic Change and Innovation in Port Organization

> The Adriatic route has been reborn, linking China and Bavaria. (Autorità Portuale di Venezia 2005, I)

> The economic future of Venice has been the subject of arguments for years, but little has been achieved by this means. Many plans have been made, but few have been realized, there have been many bureaucratic delays and intersecting vetoes ... If we overcome a phase of routine in the management of port affairs, the port could become a magnet for new logistic and industrial activities ... The re-qualification of Porto Marghera in these sectors - i.e., logistics and distribution, 'modern' chemistry and shipyards – could be the driving force for the development of the whole Venetian economy ... In this way, it will be possible to contradict the pessimistic predictions of those who see Venice as transforming into a Disneyland. (Tiziano Treu, *Il Mattino di Padova*, 28 November 2007)

The 'rediscovery' of the significance of the port for Venice's future is closely tied to the renewed dynamism of the commercial sector, which since the late 1990s helped to halt and reverse the port's critical decline (see Figure 6.2).

The commercial sector – still marginal in 1994 (comprising only about 6.1 million tons, 26.9 per cent of the port's total output) – has been steadily increasing since 1995. In 2006 it reached 14.5 million tons (47 per cent of the total). Still behind in the container market on a North Adriatic scale in 1995 (about 127,000 tons; compared to 193,000 for Ravenna and 151,680 for Trieste), in 2006 the port of Venice handled 316,000 tons (with Ravenna at 162,000; Trieste 220,000; and Koper 218,000) (Port Authorities' statistics for the various years). The commercial sector of the port expanded into new areas, thus becoming the driving force behind the physical transformation of Porto Marghera. Moreover, in the last decade, the port of Venice has become one of the most important ports in the Mediterranean cruise industry.

This renewed dynamism within the commercial sector must be placed in the context of the main geo-economic trends that have characterized the Mediterranean since the early 1990s. In particular, the following factors must be recalled (Soriani 2002; Soriani and di Cesare 2005): a) an increase in commercial

Figure 6.2 Porto Marghera: port areas (photo by Stefano Soriani)

trade – between EU countries and East Asia in particular – has helped to renew the interest in the large container ship industry in this maritime area; b) again with reference to the Mediterranean, the development of new 'hub and spoke' systems facilitated the participation of medium-sized ports in the container market thanks to its new feeder services; c) the geo-economic change in Eastern Europe led to an increase in intra-Mediterranean trade (above all, in bulk cargo and semi-finished products); d) the strengthened role of the Adriatic as a maritime corridor, resulting from both the deteriorating conditions for road transport following the war in former Yugoslavia, and the increase in commercial trade between Greece and Turkey and other EU member states (particularly Germany and Austria); e) the growing market for cruises in the tourism sector has transformed the Mediterranean into an 'emerging power' in the cruise liner industry. Consequently, we can argue that the improved performance of the Venetian commercial port industry in recent years has resulted from the ability, in local terms, to territorialize the new opportunities opened up by the globalization of the economy and the maritime industry, as well as the principal geo-economic trends at work on a global scale.

Before considering the nature and modality of this process of territorialization, another extremely important external factor needs to be considered: the passing of the Italian Port Privatization Law (n. 84/1994), which opened the door to many potential organizational changes by introducing a 'landlord' model into the Italian port system organization (Musso and Benacchio 2002). This model consists of a private-public partnership, in which the Port Authority, as a public body, takes on an administrative and regulatory role, whilst a number of private firms manage the activities of the docks and ships (for instance, terminal operators and port service companies) on the basis of rules for private competition. The passing of this law has facilitated the evolution of more complex organizational models in the port sector, together with the great technological and organizational changes that have transformed the very nature of port activity as well as its role in the global reshaping of production and distribution systems (Heaver 1995; Suykens and van den Voorde 1998).

On the one hand, the adoption of this type of model obliged the individual port actors to compete in the market, thus improving their own industrial performance; on the other hand, it forced the public body (the Port Authority) to pay even more attention to strategic tasks (that is, strategic planning, marketing policies, 'foreign policy' and coalition building). Consequently, port authorities have greatly enlarged the range of their activities in the last few years. Increasingly they contribute to such activities as environmental management, the design of transport networks, and planning at both local and regional levels. Additionally, they have acted as a kind of 'ambassador' to local and regional networks of interest, fostering and promoting land-development projects and contributing to the construction of growth-and-development coalitions on local, regional and national scales. As a result, the port authorities have become extremely important agents in the processes of both setting up and consolidating new networks which involve different types of businesses and actors on different spatial scales. Similarly, port companies (above all, the terminals) have also been subject to radical change. The main trend is for the gradual integration of port terminals into the logistic chain via vertical integration strategies, although numerous horizontal integration strategies have also been implemented (terminals in different ports being purchased by multinational companies). Consequent to these various factors, the local terminal has been able to become a node in new regional, national and even global networks. Significantly, participation in these networks can be an essential factor in competition on a local scale, which in turn, may influence the capacity to mobilize organizational and financial resources on different scales (Soriani 2002).

The Local Territorialization of Emerging Opportunities: The Dialectic Move between the 'Material' and the 'Social' and the Role of Cooperation

> The Port of Venice has been waiting for 15 years for a solution to be given to the problem of dredging … A real obstruction which has devalued the image of the Port of Venice in Italy and abroad. (Autorità Portuale di Venezia 1999, III)

> Public opinion's awareness of how important the port is for the city has grown. Equally, so has the willingness of the Port Authority to collaborate towards finding solutions of important problems for the city. (Autorità Portuale di Venezia 1999, V)

> With reference to the go-ahead for the dredging operation, ... a fundamental role needs to be recognized with perfect coordination among all interested parties. They had created a team – a difficult achievement in Venice – and it won them the port. (Autorità Portuale di Venezia, News and Sailing List 02/2006, 25)

Up until the first half of the 1990s, policies regarding the Port of Venice came under the auspices of the Provveditorato al Porto (Port Administration Offices) and the Compagnia dei Lavoratori Portuali (Port Workers Association). The former body, established in 1929, mainly provided technical and infrastructural knowledge that didn't help the promotion of organizational innovation and new approaches in the market. The latter, established in 1945, remained a cornerstone of the labour culture, bound to traditional labour-intensive operations on 'multipurpose ships' (Ernesti 2001; Federici 1996). The passing of the Port Privatization Bill led to the establishment on 1 January 1995 of the Autorità Portuale di Venezia (Port Authority of Venice, henceforth abbreviated APV) and in the (contextual) setting up of the process that brought about the formation of a group of private terminal operating companies (seven of which were still active in 2007).

Besides improving the efficiency and competitiveness of the port's activities through the privatization of wharf operations and services, APV was called upon right from the start to improve the port's infrastructure and layout, to enhance maritime accessibility via new dredging operations, to encourage consensus and trust in the port's environment, and to promote the social legitimacy of the port on an urban level. Two projects have been particularly important in this process: a) the expansion of the commercial port into areas of the Porto Marghera industrial zone, which had already been abandoned or were in the process of being abandoned; b) the initiation of dredging and maintenance operations in the port channels.[2]

---

2 The 'reconstruction' of these two processes is based on the analysis and critical interpretation of: a) the main documents compiled by the Port Authority of Venice (especially the Operative Tri-annual Plan) and the main publications edited by the above (in particular News and Sailing Lists) for the period of July 1996–May 2007; b) the main preparatory documents for the Variante al Piano Regolatore Generale per Porto Marghera (Modification of the Regulatory General Plan for Porto Marghera) (Comune di Venezia 1996); c) the main technical preparatory documents for the Ministerial Decree for the Environment on October 25, 1999, n.471 (Environmental Reclamation Regulation) and for the subsequent law on December 9, 1998, n.426 (New Interventions in Environmental Fields); d) the Accordo per la Chimica a Porto Marghera (Agreement on Chemical Activities at Porto Marghera) of October 21, 1999 acknowledged by the 'DPCM' on February 12, 1999; e) Regione del Veneto, Comune di Venezia (2004).

The Commercial Port Expansion Process

It should be remembered that those in the port sector had frequently expressed the need to expand the port for commercial activity, but for a long time it seemed unlikely to happen. With the recognition of potential opportunities for the port environment (led by the APV), and growing support for the port industry in political and cultural sectors of Venice, new and diverse interests concerning the port successfully came together. As a consequence, new social networks that re-interpreted the port's interests were created. Private terminal operators and transport workers (forwarding and maritime agents, carriers, tug boat operators and helmsmen, etc.) found that the APV provided a voice capable of expressing a new joint strategy for the port's development. At the same time, the interests of the city – its administration and its important urban elite – began to converge in a very significant way. Expansion was recurrently interpreted and described (in technical reports, local debates, etc.), as a potential revitalization tool for Porto Marghera. Moreover, as significant figures in the political world and Venetian civil society argued, the concentration of commercial activity at Porto Marghera and its improved spatial organization would actually allow the city to begin a process of partial urban redevelopment in the 'old' port, located in the western part of the lagoon city (Stazione Marittima), which until then was still being used for commercial port activities.

This convergence of interests and the new social networks (inside the port business and between the port business and the local milieu of Venice), which were set up and consolidated around them, were also essential for re-thinking the problem of redundant industrial areas. Long considered to be emblems of industrial decline, these areas began to be interpreted, represented and promoted as potential resources thanks to the dynamism of the port commercial activities.

The 'climate of waiting' that had characterized earlier attempts to find a solution to the port's problems and for the possible beginning of a new material transformation both for Porto Marghera and Stazione Marittima came to an end with an agreement between the Municipality of Venice and the APV (Protocol Agreement – Council of Venice and Port Authority of Venice on 17 April 1997). It approved the project and the realization of the so called *insula portuale* (port island). This project proposed a number of things: the expansion of the commercial port into disused industrial areas at Porto Marghera; the abandonment, for shipping purposes, of some areas of Stazione Marittima and their urban redevelopment; and the related reorganization (revised layout and improvement of infrastructure) of businesses in favour of passenger services (such as cruises) in the Marittima area.

Furthermore, the project quickly found support outside the local milieu. For many operators, the implementation of the privatization law prompted a process of progressive integration of distribution and logistic chains on regional, national and global levels. All in all, the port of Venice began to represent an increasing number interests that could no longer be circumscribed on a local level. This favoured and promoted the formation of multi-level alliances capable of re-routing new economic resources towards the development of what was emerging, both materially and socially, as the 'new port'.

Dredging and Marine Accessibility

Many planning documents and statements of the Provveditorato al Porto (Port Administration Offices) and of the Ente Zona Industriale (Industrial Zone Committee) from the late 1980s and early 1990s tried to mobilize social and political actors to pay attention to the importance of the 'dredging issue', but with little success. The dynamism of commercial port in the mid-1990s contributed to a different way of interpreting and representing this basic problem. A 'call to arms' – to dredge the port canals and improve nautical accessibility – became a central point around which new interests converged, thus strengthening the formation of new alliances that were being established. Regarding this, the discovery that port canal sediment was contaminated supported the case of the APV and the port industry as a whole. This extremely toxic canal sediment was found to be spread throughout the ecosystem, owing to processes of erosion and by the passage of ships (Magistrato alle Acque, Autorità Portuale di Venezia 1999). Following a series of environmental investigations, the City Council and the Mayor issued an ordinance forbidding fishing in the central part of the lagoon (Ordinance of the Council on the Environment, Venice Municipal Council, 9 December 1996). So yet again, the solution to the port's problems through the revitalization process of Porto Marghera (excavating the canals and to improve maritime accessibility) converged with a solution for the city's problem (improving the environment and reducing sanitary problems through environmental reclamation operations). The complexity of this scheme (in financial and technological terms) also favoured the promotion and consolidation of new cooperative relations between the city administration, the APV and the Venetian Regional Council. Furthermore, this process was accompanied by a series of initiatives – sandbank reconstruction, participation in research programs, and implementation of new environmental audit schemes – all of which demonstrated the importance of the APV as an environmental manager and as a mediating agent within the network (Soriani 2004).

The Birth, Consolidation and Cooperation of the New Port Business Community

These were the two most important phases in the development of the port, both in material and social terms. By producing and consolidating new alliances, the mobilization of new resources was promoted; and their effectiveness helped to advance the production of mutual recognition and trust in the port environment. Moreover, the fact that the commercial port was undergoing a radical transformation, not only from an economic and territorial but also an environmental point of view, would later advance the recovery of social legitimacy for its activities.

In little more than a decade, territorial transformations, led by the dynamism of commercial port activity, have profoundly modified the entire structure and landscape of Porto Marghera. At the same time, the nature of the port as a social agent has changed fundamentally. Attempting to represent the port today in terms of a clearly defined spatial environment and by fully enumerating the functions and social networks that sustain and animate it is an almost impossible task. The APV and port companies operate through (and give life to) networks that are ever more complex in their aims, nature and organizational style. With this in mind, it is no longer possible to refer to the port as an individual social agent. Rather, it is more appropriate to refer to the existence of a Port Business Community (PBC), which operates on a diverse scale and according to a diverse logic. First of all, the PBC is built on a system of networks of agents that are directly involved in port and port-related activities (port and maritime services, transport and distribution, logistics, etc.) according to the logic established by adopting the 'landlord model'. These networks operate on a local scale by competing with one another in attracting vessels and goods. Moreover, since the integration of port activities within the logistic chain is becoming critical in port competition, these networks also include national, international and even global agents as part of vertical and horizontal integration strategies. Secondly, even if these port networks compete in the market in order to pursue their own individual goals, they frequently cooperate in working for mutual advantages in terms of lobbying and negotiating regulations on the use of coastal space or by improving the quality and supply of common services. Thirdly, the PBC is continuously (re)built and strengthened through the formation and consolidation of networks that bring together port and non-port actors. The increased importance of cruising, recreational sailing and yachting in the Venetian context is the leading force in this process, as new networks between port actors, businesses in the tourism industry and land developers are realized. Finally, the PBC is built on the systems of networks which are promoted and consolidated by the 'political' action of the APV, an organization that has supported new forms of partnerships among those in both the private and public sectors – in the port environment and beyond and in diverse fields of business on a variety of spatial scales (business development, nautical and cruise sector, IT services, management and promotion of intermodal services, management and promotion of fairs, and land development projects).

Cooperation across these networks has been a fundamental condition for the process of material and social change that has occurred within the port activities of Venice and the port's influence on the local milieu since the mid-1990s. Regarding the port sector (the APV and terminal operators, other actors in transport, port and maritime services), this cooperation has played an essential role in advancing material transformation and the formation of new networks and trust. This has been interpreted, and at the same time operated, as a fundamental means of attaining market objectives in a much more complex competitive context than existed in the past. Positive outcomes have allowed the PBC to negotiate with other agents (collective or individual, public and private) for new development objectives, whilst the cooperation realized between the port (represented by the multiple activities coordinated by the APV) and the city administration has arguably been even more crucial. For ports in general, the nature of relations with cities (and with public and private agents and networks who interpret and represent them) has continuously been an issue of great importance (Hoyle 1999). It is no exaggeration to say that, in the light of the social and political complexity that characterizes Venice, both the nature of the port's relations with the city, and its capacity to re-interpret them, have been absolutely essential in producing and maintaining these relations. Moreover, in order to correctly understand the mobilization of major cooperation between the PBC and the city, we should note the importance of the city's new political climate after the election of philosopher Massimo Cacciari as the city's mayor. Cacciari was elected according to the Law of Direct Election for Mayors (Law n. 81, 25 March 1993), which paved the way for a period of so-called 'Leader Mayors' in Italian politics (Vandelli 1997). In the Venetian context, passing this law helped to reduce the extreme fragmentation that the local political system had suffered from previously. Moreover, by reinforcing the influence of an elected mayor in the formation of the Executive Council, this law also allowed important parts of Venetian civil society to play a greater role in the administration of the city. In this new political climate, arguments for the port and the political support for its development gained momentum as awareness of the role that the commercial sector could have in the redefinition of Porto Marghera increased amongst city administrators and the most influential urban elite.

'Narratives', Trust and Social Legitimacy

> Not everyone likes to remember that the Port of Venice boasts of a great historic, political, social, economic and commercial tradition. ... The Port of Venice, condemned without trial and without evidence by the false accusation that it caused a hydraulic upset in the lagoon, remains one of the main sources of wealth for Venice, its region – Veneto – and the Adriatic Sea. ... more than 10,000 ships in transit a year; 2,500 billion lire is the wealth produced by its companies; 1 million passengers a year; 18,000 people working there everyday; 100,000 seamen are in transit there every year. (Andrea Razzini, in Autorità Portuale di Venezia 2001, 7)

> There is this sort of refusal on the part of the city to be involved with activities concerning the sea. It is a historic error. Our fate lies in being connected to the sea, and so I cannot help but express satisfaction with this research project which incontrovertibly demonstrates the positive value that the cruise-shipping industry has

> in Venice. I ask you, truly, at this point to create a battle horse. We shall impress this over and over on your minds and assure you that we will create a team together with you. (Claudio Scarpa, Director of Association of Venetian Hoteliers, in Autorità Portuale di Venezia and Venezia Terminal Passeggeri 2006, 105)

Over the last decade, publications dedicated to the port – documented plans, technical reports from Council planning offices, and provincial and regional plans – have highlighted the profound material and social transformations which originated with, accompanied and supported the PBC's activity. Furthermore, these elements have contributed to defining a vision of future development (deeply rooted in local debates), which recognizes the 'new' port and port-related activities, along with the social networks which enliven the port business, as fundamental tools for counterbalancing the tendency towards tourist commodification. It is beyond our purposes here to consider the extent to which the PBC and local system may proceed along such a path. It almost goes without saying that rather than assuming that the future will be based on what has happened in the last decade, these elements of transformation should be evaluated by considering both how today's global maritime industry and port competition are changing and the ability of the local system to successfully cope within the context of these changes. What is more interesting in my view, is how such a vision has constituted an important factor in the formation of the PBC with respect to both the consolidation of new networks and to the PBC's legitimization within the Venetian political and social context. From this perspective, the social and economic dynamism of the new PBC, along with the dramatic changes that have occurred to Porto Marghera as a result of the port's increasing commercial activity in recent times, have strongly contributed to a new 'discourse' regarding the relationship between port development and the evolution of the city and its lagoon.

The commercial port is typically now interpreted as a fundamental asset for the city's future and as vital for reducing the dissonance between economic and ecological arguments concerning the lagoon. Two basic 'narratives' supported this process: the first emphasizes the role of the 'new port' as an economic engine within the process of regenerating Porto Marghera; and the second regards the recent dynamism of the 'new port' as a basic lever in the process of the 'ecological modernization' of Venice and its lagoon.

The 'New Port' as an Economic Engine

Turning to 'port impact' studies in order to assess the economic importance of the local port function has been essential in considering the role of the 'new port' as an economic engine. It is well known that the methodologies behind impact studies are potentially flawed and that they include a high degree of subjectivity when determining which elements are considered to be relevant port-related activities (Musso and Benacchio 2002). Regardless, in the case of Venice, these studies have had a fundamental impact on the social perception and representation of the port as it relates to the city and its main problems (above all, the tourist commodification and the difficult transition of Porto Marghera towards new economic functions).

Two surveys were of basic importance. The first one, commissioned by the APV (COSES 1998), offered a snapshot of the great changes in the port over a period of 20 years and at the same time confirmed that, despite the fundamental processes of restructuring over the course of the previous two decades, port activities remained a fundamental engine in the Venetian economy.[3] This investigation foregrounded the profound changes that had occurred since the Port Privatization Law came into effect, and offered a morally positive evaluation of the ongoing changes by highlighting their importance in the process of Porto Marghera's transformation. The survey constituted an important moment in the life of the PBC and helped to encourage mutual recognition and trust among port agents. The results were later used as the basis of a promotional campaign aimed at communicating how, and to what extent, the port had changed in material and social terms. In all, the survey was important both for encouraging consolidation of the newly formed alliances and for increasing the social legitimacy needed to carry out these changes.

The second investigation, commissioned by the APV and Venezia Terminal Passeggeri (the Venice Passenger Terminal), was conducted in 2005 with the aim of demonstrating the cruise industry's contribution to the Venetian economy. It concentrated on the economic impact of cruise tourists in Venice while omitting their effects on the port's own economy (Soriani and di Cesare 2005). The results confirm the economic importance of this sector for the city. Furthermore, the investigation underlined the positive implications of this sector for the development of new services and businesses in the lagoon city both in the past and the future. Moreover, this study influenced the on-going quarrel regarding the maintenance of the 'cruise port' in Stazione Marittima. The most debated questions concern navigation risks associated with ever larger ships traveling through the San Marco Basin, and the possible impact of these ships in terms of wave motion and air pollution. This debate has garnered authoritative support for the relocation of the 'cruise port' from the Stazione Marittima to Porto Marghera or further out into the open sea (at the Lido inlet and along the Cavallino coast). These options have also received support from various citizen pressure groups that have sprouted up in recent years in those areas of the city more affected by the passage and anchoring of these types of ships. These debates illustrate that despite the fundamental changes that have taken place, port activity in the lagoon city still gives rise to significant social and political conflict. It is against this backdrop that the investigation revealed the complex issues at stake in and around the lagoon. Completed in 2005, this study was published and entered public debate in February 2006. Within this debate, not

3 The data cited by Razzini (General Secretary of the APV, 1996–2004) during the period reported at the start of this section was produced by the survey in question.

only is the study referenced, but so are the principal participants in the port-tourism network who seek recognition and legitimacy in the social arena (e.g., the Venetian Hoteliers Association, the Venetian Retail Association and the Association of Maritime Agents and Shipping) (Autorità Portuale di Venezia and Venezia Terminal Passeggeri 2006). All in all, this investigation exposed a kind of 'community in movement' and helped strengthen the cohesion among and visibility of PBC members. By recording evaluations, data and forecasts, this second investigation set a precedent in terms of defensive strategy for the port environment.

Towards a New 'Partnership for the Environment'

Narratives on 'ecological modernization' (Benton and Short 1999; Reitan 1998) promulgated through the media in the form of reports, technical documents and interventions, have also contributed to the PBC's vision of a concrete remodeling of the lagoon. Through the actions of the APV, the Port Business Community has emphasized that the port's most important economic activities derive from more environmentally friendly market segments (Soriani 2004). The promotion of this vision has greatly helped in building political, social and cultural legitimacy for port activities in general and also the port-related social networks of Venice. The general belief that the PBC is, above all, developing those sectors perceived as more ecologically friendly has also helped to bridge the gap between the Venetian environmentalist movement and the economic, political and entrepreneurial spheres gravitating towards the port business. In essence, the narrative of ecological modernization has transformed 'distrust' and lack of support for the Porto Marghera project into an important facet of legitimacy for the new PBC's action.

Conclusions: Placing Cooperation and Social Capital in Context

As I have argued, the creation and consolidation of the PBC resulted from a complex process of territorialization carried out during a period marked both by dynamic external trends and by material and social transformations in the local milieu, which opened up a whole spectrum of opportunities. Significantly, a vision of Porto Marghera as a key player in the future development of Venice was concurrently elaborated and promoted, despite the progressive decline of its large industries. And indeed, port activities have become increasingly vital for the commercial sector, whilst also providing a growth area for tourism, through the cruise industry. It is not surprising then, that the PBC has promoted, and still promotes, port activities as fundamental not only for Porto Marghera's regeneration and the associated urban development, but also for the rest of the economy.

The case of Venice thus provides an excellent example of the complexity of social network formation in the processes of globalization where, on the one hand, recent trends behind globalization have favoured the formation of more complex networks, which are much harder to map out and represent on a local scale. On the other hand, 'local responsibility' is increasingly important for designing and (re)building social networks – networks that in turn are a precondition for an increased capacity to mobilize resources and accomplish projects. The case of the port of Venice provides ample evidence of the latter trend. The importance of cooperation has been demonstrated here through the way that certain actors have taken 'local responsibility' for latching onto emerging opportunities and translating them into mobilized resources that ultimately bring about territorial transformation. Cooperation has also played an essential role in building new networks in Venice, which are able to further mobilize new resources, as well as promote and sustain trust and social legitimacy.

It might be argued then that the search for greater cooperation – with regards to both the port and the city – has been strongly encouraged both by the politically and socially defensive conditions under which the port has been (re)constructed. Furthermore, this cooperation has been advanced by the contextual awareness of the urban elite and city administrative bodies as they realize the benefits that a 'new' commercial port and port-related activities could bring in terms of initiating and strengthening the material and social transformation of Porto Marghera. As I have tried to show, for the port, cooperation – both within the port environment and in terms of city planning – was increasingly interpreted and represented as a basic condition for improving performance in an increasingly complex market as well as for promoting acceptability and social legitimacy in the urban context, both of which had been profoundly eroded during past decades. For the city, cooperating with the emerging PBC was an essential step forward, both for starting the process of regenerating Porto Marghera and for commencing the waterfront redevelopment at Stazione Marittima.

I argue that in this case, attempts made by actors to find and realize mutual benefit have played a fundamental role in the establishment and consolidation of cooperation, and eventually of new social networks. Indeed, cooperation amongst social actors, within the port environment and between port actors and the city, was essential for the social and material transformation in Venice *via* the implementation of projects aimed at recontextualizing Porto Marghera and its role in the changing global economy. Through cooperation, new social alliances were set up and consolidated. In the context of Venice then, the formation and consolidation of cooperative networks can be seen to have opened up a new way of conceptualizing and practicing 'power' – shifting from a regulatory system based on having *power over* someone/something, to a system oriented towards *power to* realize certain objectives (Desfor and Keil 1999). Taking into consideration the complexity of the Venice milieu from a social and political perspective, cooperation has proven to be the key element in promoting, establishing and eventually consolidating new social networks that are best able to affect social and material change.

Though it is clearly established that cooperation was extremely important in attaining goals and producing material and social transformation, representing and evaluating the basic incentives and driving factors by means of which the 'rationale for cooperation' and the 'acknowledgement of the importance of cooperation' were translated into practice – into concrete cooperative action – is a much more complex matter. With this in mind, the Venice experience reveals how cooperation, and therefore the mobilization of resources, was not the product of something that already existed in the 'local;' rather, it highlights how cooperation and the mobilization of resources were the product of complex material and social transformation processes which redefined and recontextualized the very nature of the 'local'. Indeed, I have argued that the creation and consolidation of cooperative networks was the result of a process of 'territorial innovation' involving the following fundamental factors:

1 *Discontinuity with the past* was produced by interactions between external factors opening up opportunities, and by those within the local milieu being able to recognize such opportunities and recontextualize the milieu in relation to them. It certainly was not some ripe and ready product that had magically appeared 'as part of the wave of history'. We must not forget that, from the late 1960s onwards, Venice was a paradigmatic case for the crisis of Fordist industry in Italy. Characterized as it was by profound social conflict and the collapse of networks established during the peak years of industrial modernity in the 1950s and 1960s, the local context was substantially 'blocked' by its internal distrust and by a profound crisis of social legitimacy. In this situation it was possible to start a 'cooperative phase' and build new networks capable of mobilizing resources thanks to various external factors and events (the port privatization law, the new law for the election of mayors, the presence in the political and economic scene of a new urban elite). These were not however the direct result of an existing 'stock of trust', brought about by past developments and already present within the local milieu as a preconstituted patrimony. They had to be created through cooperative social arrangements.
2 A process of *territorial innovation* that through cooperation contributed to the creation of new trust and the establishment of a cumulative process of strengthening and legitimizing the nature and pace of material and social transformations. Such an approach to the theme of cooperation and its importance in mobilizing resources seems to be completely consistent with the developmental process that has taken place in Porto Marghera.

3 *A 'narrative'* of the economic importance of the 'new port' played an important role in the process of counterbalancing the tourist commodification of the Venice area. The 'new port narrative' has recontextualized the relationship between the port and the city, and in the process it has been able to relieve the tension between those with economic and those with ecological concerns for the future of the lagoon.

'Narratives' in general must be based upon achievements in order for them to be believable. The persuasive force of 'narratives' – and their part in sustaining the social legitimacy of networks based upon them, whose activities reinforce them – is strongly dependent upon a *previously demonstrated capability* of mobilizing resources and putting them into action for the realization of economic and social goals. At the same time, however, their fundamental function is to define and consolidate new systems of social evaluation through mutually validating attitudes and perceptions.

The publication of empirical data concerning material and social change has itself influenced the 'new port narrative', and thereby reinforcing the interpretation that previous achievements (e.g., an increase in the productivity of the commercial port resulting from the privatization process, a port expansion project and port channel dredging having begun, and the progressive consolidation of alliances which were set up for the realization of the above) were but early phases in a long process of transformation. This, in turn, advances the creation of new trust and new social legitimization. This case thus substantially confirms how 'narratives' can 'bring about' the future by sustaining the hidden mobilization of resources. In this regard, the case of Venice also helps to illustrate how trust and distrust are not determinant variables, but dynamic factors that can contribute to the recontextualizion of the 'local' by promoting innovation in the way it represents itself. The phase of distrust and crisis in social legitimacy that characterized Porto Marghera in the 1970s and 1980s has constituted a fundamental factor in the formation of new networks, new trust, new social legitimization and eventually, social capital. Environmentalism and radical criticism of the Fordist experience in the Lagoon during the 1980s have proven themselves to be fundamental factors for socially, politically and culturally 'validating' the action of the new PBC and the vision of future development it promotes.

Bibliography

Autorità Portuale di Venezia (ed.) (2001), *Venezia, Guida al Porto* (Venice: Marsilio).

Autorità Portuale di Venezia (2002), *Piano Operativo Portuale 2000–2002* (Venice: Autorità Portuale di Venezia).

Autorità Portuale di Venezia and Venezia Terminal Passeggeri (2006), *L'importanza della crocieristica per Venezia. Ricadute socioeconomiche del traffico passeggeri sul territorio* (Venice: Venezia Terminal Passeggeri and Autorità Portuale di Venezia).
Barbini, E. (ed.) (2002), *Cantiere Venezia. Piani, progetti, realizzazioni, imprese* (Venice: Marsilio).
Benton, L.M. and Short, J.R. (1999), *Environmental Discourses and Practices* (Oxford: Blackwell).
Bourdieu, P. (1986), 'The forms of capital', in J.G. Richardson (ed.).
Bourdieu, P. (1998), *Acts of Resistance: Against the Tyranny of the Market* (New York: New Press).
Comune di Venezia (1996), *La pianificazione urbanistica come strumento di politica industriale: la variante al prg per Porto Marghera* (Rome: Urbanistica Quaderni).
COSES (ed.) (1998), *Analisi della funzione portuale veneziana*, Rapporto di ricerca n. 20, May (Venice: COSES).
Costa, P. (1988), 'Venezia possibile, Venezia probabile, Venezia certa: Il turismo', *Idea di Venezia, Quaderni della Fondazione Istituto Gramsci Veneto* 3–4, 45–59.
Costa, P. (1993), *Venezia. Economia e analisi urbana* (Milan: ETAS).
D'Agostino, R. (2002), 'Piani e progetti per nuovi scenari urbani', in E. Barbini (ed.).
D'Agostino, R. (ed.) (2004), *Piano Strategico di Venezia. Linee, Strategie, Politiche* (Venice: Comune di Venezia).
De Rita, G. (1993), *Una città speciale, rapporto su Venezia* (Venice: Marsilio).
Desfor, G. and Keil, R. (1999), 'Contested and polluted terrain', *Local Environment* 4:3, 331–52.
Ernesti, G. (2001), 'Venezia da terra, Venezia da mar. Città e porto nell'età dell'industria', in Autorità Portuale di Venezia (ed.).
Federici, G. (1996), *Portuali a Venezia* (Venice: Il Cardo).
Governa, F. (1997), *Il milieu urbano. L'identità territoriale nei processi di sviluppo* (Milan: Angeli).
Heaver, T.D. (1995), 'The implications of increased competition among ports for port policy and management', *Maritime Policy and Management* 22:2, 125–33.
Hoyle, B.S. (ed.) (1996), *Cityports, Coastal Zones and Regional Change: International Perspectives on Planning and Management* (Chichester: John Wiley and Sons).
Hoyle, B.S. (1999), 'Scale and sustainability: the role of community groups in Canadian port-city waterfront change', *Journal of Transport Geography* 7:1, 65–78.
Indovina, F. (ed.) (2003), *Progetto Metis. Sui possibili scenari futuri di Venezia e della sua Laguna* (Venice: CORILA).

Magistrato alle Acque, Autorità Portuale di Venezia (1999), *Rapporto sullo stato di compromissione dei fondali* (Venice: Autorità Portuale di Venezia).

Mannino, I., Soriani, S. and Zanini, F. (2004), 'Bonifiche ambientali e riqualificazione delle aree portuali-industriali. I casi di Anversa e Porto Marghera', *Rivista Geografica Italiana* 111:2, 201–31.

Musso, E. and Benacchio, M. (2002), 'Demaritimisation o remaritimisation? L'evoluzione dello scenario economico nelle città portuali', in S. Soriani (ed.).

Musu, I. (ed.) (1998), *Venezia sostenibile. Suggestioni dal futuro* (Bologna: il Mulino).

Putnam, R. (1993), *Making Democracy Work: Civic Traditions in Modern Italy* (Princeton, NJ: Princeton University Press).

Regione del Veneto, Comune di Venezia (2004), *Master Plan per la bonifica dei siti inquinati di Porto Marghera* (Venice: Regione del Veneto and Comune di Venezia).

Reitan, M. (1998), 'Ecological modernisation and "realpolitik": Ideas, interests and institutions', *Environmental Politics* 7:2, 1–26.

Richardson, J.G. (ed.) (1986), *The Handbook of Theory: Research for the Sociology of Education* (New York: Greenwood Press).

Rinaldo, A. (1997), *Progetto Venezia 21: Equilibrio fisico e idrogeologico della Laguna.* Rapporto di Ricerca 09.97 (Venice: Fondazione ENI Enrico Mattei).

Slack, B. and Pinder, D. (eds) (2004), *Ports in the Twenty-first Century* (London: Routledge).

Soriani, S. (1996), 'The Venice port and industrial area in a context of regional change', in B.S. Hoyle (ed.).

Soriani, S. (ed.) (2002), *Porti, città e territorio costiero. Le dinamiche della sostenibilità* (Bologna: il Mulino).

Soriani, S. (2002), 'La transizione postindustriale della portualità tra dinamiche di mercato e vincoli-opportunità territoriali', in S. Soriani (ed.).

Soriani, S. (2004), 'Port development and implementation challenges in environmental management. The case of Venice', in B. Slack and D. Pinder (eds).

Soriani, S. and di Cesare, F. (eds) (2005), *L'importanza della crocieristica per Venezia*, Centro IDEAS, Research Report, November (Venice: Università Ca' Foscari).

Soriani, S. and Zanetto, G. (2002), 'Portualità e dinamiche sostenibili. Il caso veneziano', in S. Soriani (ed.).

Suykens, F. and van de Voorde, E. (1998), 'A quarter of a century of port management in Europe: Objectives and tools', *Maritime Policy and Management* 25:3, 251–61.

Vandelli, L. (1997), *Sindaci e miti* (Bologna: il Mulino).

Zanetto, G. and Tassan, F. (1990), 'L'industrializzazione costiera e i processi di regionalizzazione litoranea', in COSES (ed.).

PART 3
Mobilizing Trust

Chapter 7
Speaking Triestino: Language, Practice and Social Capital in Trieste

Claudio Minca

> In Trieste, there were the Greeks, the Armenians and the Turks, the Sicilians and the Maltese, Germans, Slavs and Hungarians …; along with the Sicilians, there were also the Italians from other regions; and the Egyptians and the British too. You would think that Trieste, just as the souk of some Oriental city, was a cacophony of sounds, a cross-roads of twenty-odd languages, a modern-day Tower of Babel where, unlike in the case of the latter, everyone understood everyone else. But this was simply not the case: *everyone spoke triestino.*
>
> Alberto Spaini, Autoritratto Triestino 1963, 29; italics added

Introduction

In one of his most recent books, *Trieste Sottosopra* (2006), Triestine novelist Mauro Covacich analyses the relationship between language, practice and place, while watching local teenagers diving off the shore of Barcola – Trieste's popular urban beach (Figure 7.1). Covacich wonders how is it possible that, generation after generation, the dives don't change – and carry the very same names in dialect, untranslatable into Italian but wonderfully evocative of the different types of performance, staged for an always-appreciative audience. The art of the perfect dive is not necessarily something passed on from fathers to sons, and neither are the colourful descriptions that capture these performances. Covacich wonders, then, if it is not precisely that rocky shore, that glistening sea – that which he sees as a distinct *triestinità*,[1] translated into space on the Barcola beach – that binds the past and the present; that brings together ways of being and ways of doing that are eminently 'triestine;' a unique *koine* that manifests itself at all the key moments in the history of the city, but that also marks its most mundane, banal aspects.

It is the existence of such a sphere of 'deep' communication (if I can term it that) constituted by the use of dialect that I will attempt to interrogate here and, in particular, its role as powerful expression of a distinct substrate of unwritten rules, of shared 'ways of doing' and shared pasts; of a series of

1 I will adopt this term to signify a distinct Triestine 'way of being', a set of taken for granted understandings and practices, embodied in the use of the dialect.

Figure 7.1 Barcola beach (photo by Claudio Minca)

'principles of recognition' that make the activation of a specific form of *Triestine social capital* possible. What I will try to argue is that without the dialect – here intended in its broader sense, not only as linguistic form but also a distinct set of sedimented practices and understandings – it is difficult, if not impossible, to access Trieste's immaterial economies, to activate those dimensions of sociability often identified as 'social capital' (for a fuller discussion, see the chapters by Cannone and Giaccaria in this volume).

The formation of a group identity, according to Isin and Wood (1999, 19) is, broadly speaking, a process through which individuals recognize in each other certain attributes that establish resemblance and affinity. These attributes then become the basis of identification whereby 'individuals actively produce and reproduce *equivalent dispositions*'. As Isin and Wood argue, the dialogical process of recognition is therefore 'an ongoing negotiation of habituating, inculcating, defining, redefining and reproducing these dispositions'. This is perhaps the hidden link that Mauro Covacich locates within the waves on Barcola's beach – the *triestinità* that, he claims, binds the past and the present 'in place'.

Pierre Bourdieu engaged with these questions through his interpretative analysis of groups within three interrelated concepts: those of 'habitus', 'field' and 'capital'. Bourdieu defined habitus as a 'system of structured, structuring dispositions, which is constituted in practice and is always oriented towards practical functions' (Bourdieu 1990, 52). The formation of a group thus involves the inculcation in individuals of specific ways of being in the world – and making sense of that same world. As Isin and Wood (1999, 39) argue:

> Since habitus is a system of durable dispositions it can be seen as a presence of the past that tends to perpetuate itself with continuity and regularity. It is this regularity that objectivism records without being able to account for it. It is also this regularity that leads advocates of groups to essentialize and naturalize habitus … This is what the subjectivist approaches miss: while habitus is socially produced, its outcome is embodied in gestures, bodily movements, ways of thinking, manners and a series of very real effects.

With this understanding in mind, the *triestino* dialect will be analysed here as *a territorialized form of habitus*, while the *triestinità*, both in its materialities but also in the imaginations upon which it draws, will be conceptualized as the 'field' within which Triestine social capital takes form and gains meaning. Since the theoretical framework guiding this approach is well-developed in the initial three chapters of this book, I will concentrate here on providing an initial (and somewhat informal, given the subject matter) attempt to understand the relationship between social capital, trust and language practice in Trieste, long a formidable laboratory of European modernity (see Magris 1997; 2007; Ara and Magris 1982; also Agnelli 2005).

Figure 7.2 Piazza Unità d'Italia, the symbolic core of Triestine social life and political (photo by Claudio Minca)

Speaking Triestino

The *triestino* dialect boasts origins that are somewhat distinct from those of most Italian dialects (see De Mauro 1963). Trieste's dialect is born, in fact, as a pragmatic-commercial tool of a new cosmopolitan bourgeoisie that, beginning in the second half of the 1700s, affirms itself as the dominant class in the Adriatic port entirely re-invented by the Habsburg Monarchy. This new urban class establishes its identity precisely through the adoption of a new *lingua franca*, a variant of the so-called 'colonial Venetian' dialect which, transformed by myriad Central European, Balkanic and Mediterranean influences, will eventually take the shape of the *triestino*.

Modern Trieste is, in many ways, the product of two Imperial decrees: the proclamation on the part of Charles VI of free transit in the Adriatic in 1717 and, subsequently, the constitution of Trieste as a Free Port two years later. The crucial moment for the city comes with the birth of modern Austria: in the second half of the eighteenth century, the 'enlightened' Empress Maria Theresa transforms Trieste into the key commercial port of the Habsburg Empire (Ara and Magris 1982, 18). Trieste becomes 'the pearl of the Adriatic' and, in its new role as a hub for international trade, gives birth to a series of myths that accompany – and sustain – the city's rapid expansion and the emergence of a new cosmopolitan bourgeoisie hailing from all parts of Europe and the Mediterranean, attracted both by the extraordinary economic possibilities offered by Trieste in that period, as well as the city's climate of cultural and religious tolerance, favoured by its free port status (see Dubin 1999; Purvis and Waley forthcoming).

All through the nineteenth century, the Adriatic city will come to be considered by Vienna the key financial and bourgeois-capitalist commercial centre of the Empire, but also the first port of call along a romantic-nostalgic itinerary into 'the country where lemon trees bloom' (Rumpler 2003, 55). This imaginary will firmly impress itself upon Trieste's self-representation. Arduino Agnelli (2005, 123), in his masterful archeology of the idea of Mitteleuropa, notes how this ideology – that dominates Imperial imaginations all through the 1800s (see also Magris 1996; 2005) – finds in Trieste its ideal substrate. In particular, the dream of creating a 'new European (or, better yet, Mitteleuropean) subject', free from the shackles of petty nationalisms, modern in outlook and well-versed in questions of commerce and trade, finds here its strongest support – and clearest materialization. It is through individuals such as Baron Von Bruck, perhaps the best known and most influential promoter of the Mitteleuropean imaginary as an antidote to nation-state ideals, that Trieste begins to think of itself as the 'homeland' of a new 'industrious and cosmopolitan' bourgeois subject; a subject that, at home in Trieste, speaks nothing other than the *triestino* dialect (Cattaruzza 1995; Millo 2003; see also Jonsson 2000).

The new Trieste quickly becomes 'a key crossing point for all commercial traffics going from North to South and vice versa: the entire southern coast of Mitteleuropa is condensed into a single city' (Ara and Magris 1982, 231). In the dynamic phase that follows the proclamation of the Free Port, Trieste sheds its historical origins as a small (though always fiercely autonomous) urban centre, dominated for centuries by the very same 13 patrician families (Pellegrini 2001, 293). The memories of this urban past are not sufficient for the new Trieste, however: the booming port-city has to invent for itself a new sense of unity, a new commonality around which to rally the interests and identities of its continually changing demos. What is needed is an idea – and an identity – able to gather within it, on the one hand, the broader role played by Trieste in European and Oriental trade routes and, on the other, the direct economic interests of the city's new protagonists, that is, its commercial-economic elite (see Millo 1989). Since such an identity cannot draw on the city's past, it makes of the city's present economic vocation its guiding cultural-identificatory model as well: Trieste as the cosmopolitan, bourgeois capitalist emporium that, logically, adopts as its 'mother tongue' a variant of the main commercial idiom of the Mediterranean – the 'colonial Venetian' that for long served as *lingua franca* all along the shores of the Adriatic and wide swathes of the Mediterranean – that, translated 'on the ground' in the Triestine babel, will soon become a tongue of its own: the *triestino*. The city's traditional ladino vernacular is thus displaced by a new idiom: the Venetian tongue of maritime trade (Negrelli 1978).

As Valdevit (2004, 1) argues, the history of Triestine society through most of the 1800s is marked by a process of progressive integration, that occurs without eliminating the city's 'original' identities (that is, those of the 'historical' nations of the 1700s). A key tool of integration is Italian culture and, together with it, the adoption of the *triestino* dialect; another is Trieste's unique urban 'particularism' that, for presumed 'historical reasons', affirms a direct (and equal) relationship for the city's elites with the central state. It is within these two imaginations that Trieste's distinct urban citizenship is born: a citizenship, as Valdevit (2004, 1) suggests, that emerges out of 'integration, not assimilation, because it allows [the city's inhabitants] to maintain their original ethnic-cultural identities, especially within the religious sphere (whether Jewish, Greek-Orthodox, Armenian, or Protestant)'.

Exploring the Language of Practice

With this distinct history in mind, I will now try to trace some of the political-cultural coordinates that have made of the *triestino* – and the practices and 'ways of being' that accompany it – the privileged (if not sole) space for the enactment of a specific, hegemonic expression of Triestine social capital. This chapter will argue, indeed, that it is impossible to understand the functioning and the 'territorialization' of Triestine social capital without taking into account the crucial role played by the dialect in all of the material and, especially, immaterial exchanges based upon trust; based upon personal relationships and (largely invisible) social and economic networks that activate – sometimes successfully, sometimes not (see the chapter dedicated to Trieste's failed Expo 2008 bid in this volume) – a shared set of understandings, a shared set of ways of saying and ways of doing what needs to be done. Today, the *triestino* is still widely used in all sorts of environments – from the football field to the office; from the bakery to the spaces of memory and its various celebrations, serious and not-so-serious (in fact, jokes can *only* be told in dialect). The *triestino* continues to change, to transform itself, 'keeping up with the times'. As a spoken language, it is based on shared practices; it has no rules other than those of those who speak it *in a particular moment and a particular place*. Precisely because it is the language of practice, it continues to play a key role in defining the relationships 'that matter', promises that are maintained, that don't require written agreement (see Häkli's chapter in this volume); marking exactly the sort of 'immaterial dimension' linked to specific places and modes of communication that Putnam (1993; 2000) attempted to map (in vain) in his research on the distinct local political and productive cultures of the Italian peninsula.

The argument that follows will thus be both descriptive and analytical, having to grapple with the complexities of a topic of this sort, in many ways 'untranslatable' into text. When examining the role and place of dialect, 'empirical evidence' is of limited use, for it is oral *practice* that is the true terrain of analysis (forcing the researcher into a highly problematic attempt at 'participant observation'). It is a terrain of analysis that, by its very nature, defies classification and representation; that can be approached only through specific 'performances' – that is, by speaking of the *triestino* with the *Triestini* in *triestino*!

Recent academic literature notes, indeed, the impossibility of translating into 'data' and 'empirical evidence' the immaterial dimensions of socio-economic relationships. Along with this problem of 'translation', there are also the methodological challenges that necessarily mark every research endeavour that presumes to investigate practice, and oral practice in particular (again, impossible to reduce into mere text). It is useful to recall here, for example, the comments of Glennie and Thrift (2005), engaging a similar challenge in their recent analysis of the spatial history of time in England. Having to 'reconstruct' a series of historical practices, they argue *a priori* that theirs is a theoretical paper, fully aware of the criticisms often levelled at practice-based approaches, usually predicated upon on the idea that they only offer pseudo-explanations (2005, 165). They argue that such criticisms are based on the conviction that 'inference from common behaviour to a supposedly underlying source in shared presuppositions cannot be sustained; that the casual powers of practices often seems mysterious; and that the transmission or reproduction of practices over time, and from one practitioner to another cannot be accounted for' (ibid.). Glennie and Thrift counter such criticisms by arguing that 'the historical record can meet at least some of these objections by producing specific evidence (at least, where it is available: clearly processes like everyday talk are rarely able to be recovered)' (ibid.). If we follow this line of argument, the use and the historical meaning of the *triestino* can only be understood through recourse to historical testimony that, nonetheless, cannot 'demonstrate' its power/role in that particular moment in time – and that certainly cannot presume to 'reconstruct' it since it falls into the realm of the oral and not of the textual. The authors suggest, however, that this theoretical challenge can be met through the concept of 'communities of practice'. In their understanding, such communities of practice are shared enterprises formed over time; they are contexts of significance

> 'in which we can work our common sense through mutual engagement' (Wenger 1999, 47). These shared engagements involve the negotiation of meaning (involving talk, gesture and other forms of communication and specific dilemmas), certain levels of participation ... and the production and circulation of objects [and language codes] that confirm a community's existence and occupy much of its collective imaginary. (Glennie and Thrift 2005, 166)

These communities, Glennie and Thrift argue, have over time begun to involve circulation over greater and greater spatial extents, and have increasingly become populated with 'intermediaries' who allow circulation to occur and, at the same time, shape the character of the (common) experience. The 'community of practice' that I will interrogate in this chapter, then, are the *Triestini* in their everyday production of the *triestino* as an ever-mutating and multiform element of identity. The 'intermediary' of this community of practice is, of course, the *triestino* dialect itself: as a conduit of meaning and specific values, but also of people, projects, and imaginations.

With the above in mind, I will attempt to explore the workings of the dialect through a particular understanding of trust 'in place', keeping in mind, nonetheless, that which Isin and Wood (1999, 26) identify as the necessary distinction between hypothetical – i.e. the result of our attempts to theorize groupings – and 'real' groups: 'the former is the work of symbolic or classification struggles, while the latter is the result of the practical and political work of organising and mobilising.' The *triestino* takes on this function at a particular moment in time and continues to exercise it still today, as a site within which/through which such 'work' is (informally) organized and mobilized. As Isin and Wood (ibid.) stress, the labour of group making necessarily also involves the recognition 'that a group as such does not exist, that groups form only as a result of specific practical work, but that such work is likely to succeed only when supported by real experiences'. This, then, is the 'deeper' dimension to which the strategic use of the *triestino* appeals, today as it did over a century ago. This is possibly why the *triestino* is still commonly perceived as the 'true' language of daily experience, as the sphere of everyday communication, whose boundaries are always and everywhere negotiated. If we agree with Isin and Wood's assertion that 'group identifications and affiliations and disassociations or differentiations are always multiple, fluid and overlapping', then, the enactment of a hegemonic Triestine social capital is based also on the historically grounded capacity of its dominant members to inculcate a specific vision of the *triestinità*, of a specific Triestine way of being and of working together. We should also remember, however, that though the *triestinità* may often be presented as a harmonious narrative of practices and a specific ideal of civic life, 'any group identity does not exist as an immutable and given fact of life, but it is rather the result of the narrative strategies that value certain attributes and devaluate others' (Isin and Wood 1999, 30); this, of course, is true in the case of Trieste as well.

My considerations of some of the theoretical debates on group formation, practice and social capital end here; these debates are explicitly engaged in much greater detail by the other chapters in this volume. I will focus my attention, rather, on an analysis of the formation of a presumed 'triestine social capital' *through its dialect*. This will be a 'semi-formal' exploration, given the eminently (in)formal subject of analysis; since I will be investigating an oral tradition and everyday oral practices, my sources will be historical documents that 'recall' the use of dialect, a series of shared (mostly oral) memories that still structure the Triestine collective imaginary, and various popular practices that express the *triestinità* in the everyday (distinct ways of saying and seeing, as well as songs, jokes, and endless self-celebratory myths) (see Camuzzi 2003).

It is a rich field of analysis, albeit a somewhat 'slippery' one, difficult (if not impossible) to gather within a comprehensive conceptual framework. This complexity and 'fluidity' makes it no less relevant, however; indeed, the most important theoretical point that this chapter will attempt to advance is that research on social capital – as the 'slippery' cultural expression of (an equally slippery, in analytical terms) sociability – cannot ignore this imperfect and seemingly 'unrepresentable' dimension. Otherwise, we truly risk producing 'sociology without society' (as John Urry (2000) admonished some time back).

Beginning with a consideration of its historical roots and moving on to an examination of its contemporary manifestations, I will attempt to reconstruct those 'traces of *triestinità*' that, still today, are key to 'doing business' in Trieste, that presume an entrenched taken for granted of ways of doing and ways of being. It is a series of entrenched practices and 'local knowledges' that is often derided by the *Triestini* themselves as something quite vague, not to be taken too seriously. Yet it is within this nebulous sphere that theory and practice, business and pleasure, converge and are able, in specific instances, to act as powerful motors of economic and cultural innovation – but also equally powerful markers of inclusion and exclusion.

Lingua Franca

As I noted previously, modern Trieste develops thanks to the opening up of maritime traffic in the Adriatic and, especially, its constitution as a Free Port. Thanks, in large part, to the grand Theresian projects for the redevelopment of the city all through the 1700s, Trieste is transformed into a formidable commercial emporium that opens the heart of Europe to the Mediterranean and to the new Oriental sea routes. Trieste rapidly becomes, in fact, a mythical place, identified in the historical literature as a model of cosmopolitan municipalism and the multicultural centre of a Mitteleuropa reborn: 'the most European city in Europe, the *urbs europeissima*' (Winkler 2000, 338 in Rumpler 2003). In just a few decades, a new 'bourgeois city' emerges alongside the traditional inhabitants of Trieste: made up of Germans, Greeks, Croatians, Armenians, Jews and other Italians (Rumpler 2003, 67). This new bourgeoisie rapidly becomes a powerful force in claiming further autonomy for the Adriatic city, creating a paradoxical political situation that will persist all through the next century: the Habsburg monarchy will, in fact, never succeed in co-opting into its grand modernizing (and centralizing) vision for the city the very political and social class that the Empire itself created with the 'invention' of Trieste as Vienna's commercial emporium. This is because the new urban bourgeoisie soon becomes, for all extents and purposes, true *triestine* (Rumpler 2003, 68; see also Cattaruzza 1995; Millo 1989).

As Negrelli (1978) suggests, the emporium attracts to it a particular class of individuals: 'those believing in the ideals of free masonry, those who had suffered discrimination elsewhere on account of their different religion or creed – and to whom the Theresian legislation had conceded complete freedom of worship and association in the Adriatic city.' Already at the end of the 1700s, Trieste becomes a privileged observatory for the complex social and political realities of the Empire (see Dubin 1999). Indeed, while the Monarchy is still in the throes of a reformist Enlightenment, seeking to eliminate Mediaeval resistance to the modernizing and centralizing forces of the Imperial project, Trieste already begins to witness the first stirrings of opposition to Habsburg statism. It is within this distinct historical-political context that the *triestino* dialect first emerges, and it is a legacy that it will carry with it up to this day.

The new Trieste changes, in fact, also its language. As noted in the introduction, the city's traditional ladino vernacular is replaced by a new idiom: the Venetian-Italian dialect of maritime trade (Negrelli 1978). This Venetian-Italian dialect rapidly becomes the official language of the Trieste Stock Exchange, as the city affirms itself as a new locus for maritime trade between the Adriatic and the Levant, where the Venetian dialect served, for centuries, as the *lingua franca*. In the multinational Triestine cauldron of the boom years, a particular variant of the 'colonial Venetian' dialect takes hold, also thanks to the important role in the city of Greek and other 'Levantine' populations that, 'back home' had already become accustomed to the use of the Venetian dialect as the language of commerce and trade (see Cusin 1930, 113). Historian Roberto Finzi (2001a) argues, indeed, that a clear distinction must be made between 'proper Italian' and the Trieste's *lingua franca*, a variant of the colonial dialect spoken in countless Mediterranean ports in the nineteenth century that had as its base Italian and other Romance languages, but that also incorporated Arabic influences. To understand the linguistic and identitary destiny of Trieste, it is crucial to keep in mind the role played in the formation of the modern, mercantile city by this particular variant of Italian as a *lingua franca*. The Venetian-Italian dialect is appropriated by the Triestine bourgeoisie as a key modernizing tool. The *triestino* becomes the linguistic and cultural terrain of choice for Trieste's new mercantile and financial class, which adopts it as a means of communication not only amongst themselves but also with other parts of the Mediterranean and some members of the Imperial aristocracy as well (Finzi 2003, 324). So it is for historical and economic reasons that this Adriatic Italian dialect becomes the hegemonic language in the Triestine melting-pot, while the countryside that surrounds it continues to speak largely Slovenian, and the Imperial bureaucracy, German (Merku 2001, 294; on the history of Slovene triestines see also Kacin-Wohinz and Pirjevec 1998; Verginella 2001).

This cultural-linguistic hegemony does not, however, prejudices relations in other spheres: modern Trieste's 'creation' by the Habsburg Empire as a multinational, multi-confessional free port, produces a context favourable to economic collaboration between the city's diverse faiths and ethnicities without, at the same time, dissolving the distinct group identities of the different communities (Catalan 2001, 483). Trieste's large Jewish community, in particular, becomes one of the most important forces in the city's economic and commercial renewal, and plays a key role in the formation of a modern urban bourgeoisie, marked by a unique, cosmopolitan imprint (Ara and Magris 1982, 34–5; see also Cervani 2006). The personal and religious freedoms guaranteed by Trieste's status act, on the one hand, helped to consolidate the city's Jewish community, allowing it to gain a unity that it previously lacked, given its varied immigrant origins (Trieste's Jews hailed from all corners of Europe and the Mediterranean); on the other, this consolidation of Jewish specificity becomes just that, a cultural-religious difference that remains always secondary with respect to the common values and ideals binding most *Triestini*, irrespective of language or faith (ibid., 35). The language adopted by the Jewish merchants and traders is, therefore, the city's dominant Adriatic-Italian dialect, a dialect in part already familiar to the Levantine Jews, as well as those hailing from the Balkans and Dalmatia (Catalan 2001, 483).

In this process of integration of the city's various new arrivals, language is thus of vital importance: 'Italian [actually, its Triestine version] is the traditional language of the city, but also the language most widely spoken both in the Julian hinterland as well as the Adriatic and Mediterranean ports from which most of the newcomers hail' (Ara and Magris 1982, 34). Ara and Magris insist (ibid.) on how the *triestino* dialect plays a key role here: 'diffuse among all the social classes, the vehicular language for all family and professional exchanges; mastery of the dialect is the mark of full integration into the urban culture, the necessary condition of indigeneity.'

As Roberto Finzi (2001a, 317–19) suggests, the language in which the 'modern' *Triestini* communicate, the language that determines and marks their distinct identity within the Empire, cannot simply be deduced from their ethnic or cultural origins or national-political choice; nor is it a 'logical' consequence of the Italian origins of the city and its citizens, as some historians have claimed (see Tamaro 1974). The choice of speaking the dialect is an *ideological* choice: the *triestino* is a marker of a certain (urban) identity and a certain set of (urban) values. It becomes the public *lingua franca* of a dynamic new social class – not the private space of refuge in 'tradition' and the 'local', as was often the case for other dialects (Pellegrini 2001, 293). As Doria (1978) argues, this new urban identity succeeds in bringing together the most disparate communities and individuals, a 'motley crew' of adventurers, merchants and traders now granted a new 'moral' belonging. The dialect plays a key role in this process, soon becoming a key marker of distinction in the emporium where everyone spoke *triestino* (Pellegrini 2001).

Triestinità

Although Vienna continues to formally sustain the city's economic development and its political autonomy all through the 1800s, the Imperial capital begins to increasingly question Trieste's 'Italianness'. The centralizing politics of Joseph II, coupled with new Imperial edicts destined to sustain national minorities, act to progressively transform Triestine 'municipalism' (that never questioned, nonetheless, the city's belonging to the Imperial project) into an Adriatic irredentism; an irrendentism that will take on increasingly harsh tones towards the end of the century, with its progressive *rapprochement* with the Italian irredentist movement (Rumpler 2003, 57).

The dialect – as an 'Italian' cultural space, and thus an alternative to the German one – soon becomes also the vehicle of a unique sort of municipal patriotism. Maintaining linguistic uniqueness becomes a political platform and the instrument for the affirmation of a local particularism that opposes the Empire's attempts at centralization. It is in this period that the idea of the 'Trieste nazione' (a Triestine 'homeland') first emerges, theorized and sustained by the prominent urban historian Domenico Rossetti (1815) as the formal ideology of the *triestinità*.

In Rossetti's vision, the Triestine bourgeoisie finally finds the sort of unifying myth able to gather within it the various contradictions of Triestine culture and able to provide a common ideology, a common model of urban civility. Faced with the increasingly centralizing (and Germanizing) politics of the Imperial bureaucracy, the multinational elite of this Adriatic port opts for a 'Triestine patriotism', a patriotism that will express itself in *triestino*. Rossetti's ideal of the '*Trieste Nazione*' captures this feeling of distinction very well, bringing together the city's diverse communities and subjects under a common identificatory umbrella (see Minca forthcoming). It is also an ideal that successfully engages the city's *self-image* as a plurinational, cosmopolitan, modern bourgeois metropolis; it is an ideal that captures the popular imagination that will then reproduce it – and recount it – in the everyday idiom, in that *triestino* that had by then become an extraordinary element of unification (Negrelli 1978). In Trieste, just as in other parts of the Empire, the perception of 'national' distinction first passes through the identification of a linguistic and cultural specificity. Since the dominant element both in the city and among its governing class is Italian (or at least a certain form of 'Italianness'), it is Italian cultural nationalism that increasingly begins to affirm itself in opposition to Imperial politics. The *triestino* will again play a key role in this process, as the conduit and integrating force of a very specific, Triestine and Adriatic, interpretation of 'Italianness'. This distinction is important, for the sense of cultural distinction that emerges in the city has very little to do with the sort of political nationalisms that begin to materialize across Europe in those years.

Indeed, alongside an Italian 'cultural nationalism', what emerges is a distinct 'Triestine patriotism': an ideal that 'finds its concrete expression in the everyday practices of the cosmopolitan emporium, in its distinct urban culture, customs, and its specific way of conceiving social and economic relationships ... in the culture of the plurinational Triestine society that finally finds within it a way of expressing its uniqueness, its distinctiveness' (Negrelli 1978, 382; see also Minca forthcoming). This sense of unity is not fictitious: both on the political Left and Right, the urban elite emerges from the very same cosmopolitan koine, from the very same governing class – the merchant-financial one – tightly bound by visible and not-so-visible ties and networks, from economic dealings to Free-Masonry (Negrelli 1978, 120). After 1848, Triestine patriotism reaches its apex, with its most ardent supporters proclaiming, albeit only for a brief while, the foundation of the 'Republic of San Giusto'[2] in the city. In those years, Trieste's City Council and its Chamber of Commerce become very powerful actors in the negotiation of relations with Vienna, increasingly articulated within a liberal-nationalist Italian (albeit in its Triestine variant) agenda (Rumpler 2003, 78).

Cervani (2006) suggests that the urban imaginations that characterize the early 1800s are, indeed, an idyllic extension of the eighteenth-century world that still ignores the romantic nationalisms of the nineteenth. Karlsen and Spadaro (2006, 135), in reflecting upon the distinct nature of Triestine cosmopolitanism, similarly argue that 'national' ideologies took on a very particular tinge in the Adriatic emporium: here, 'national' identification 'took on the form of a moral ferment, the growing sense of a 'spiritual belonging' to a grand, cosmopolitan

2 San Giusto is the patron saint of the city.

commercial emporium'. It is, more than anything else, a form of class distinction – a belonging to a certain modern urban elite of merchants and traders that 'make of their vocation an [urban] identity' (2005, 128). In Trieste, all financial, commercial and political relations are articulated in *triestino*, the language of the urban elite. This renders it the ideal terrain for the development of a distinct hegemonic form of social capital that soon becomes the key motor of the economic development and prosperity of the city (together, of course, with the still quite favourable, albeit increasingly tense, relationship with Vienna). As Trieste becomes the second largest port in the Mediterranean and the seventh largest in the world, what its successful elite expresses (also through its shared language) is a sort of 'illuminated conservatism', a distinct dominant urban ideology drawing upon that very same immaterial dimension that produces, on the ground, what we would call today 'networking' and 'trust' – that distinct 'sense of togetherness' that the debates on social capital have been focussing on for the last decade or so (see Millo 2001; 2003).

In his research on the formation of the emergent hegemonic Triestine social capital, Gatti (2001, 359) notes the very specific characteristics of the 'New Triestine Man': an eminently bourgeois figure, concentrating within it 'all the essential qualities of the Mitteleuropean spirit', in the words of Baron von Bruck – at that time a key player in the city's political scene. In what Triestine historian Elio Apih (1988) described as 'Austria's most bourgeois city', social capital was built upon a very specific 'national model' that acted as a key factor of urban integration. The Triestine (urban) 'nations' were much more than simple coagulations of ethno-religious difference, however. The city's 'nations' and its particular brand of (urban) cosmopolitanism were two fundamental traits of Triestine society from the late 1700s onwards (Gatti 2001, 376) and the social capital that made possible the city's rapid development was activated within commercial and financial networks bound within a greater (urban) project organized precisely around the city's 'cultural nations' and class distinctions. As Gatti himself argues, in Trieste 'nation and social class did not have rigidly defined boundaries' (2001, 376): they simply represented, respectively, forms of cultural and economic distinction that were always in a dialectical relationship.

The internal cohesion and the degree of interdependence among the most important urban economic actors was highlighted by Farnesin in his network analysis of the key economic and political institutions of the day (including the Chamber of Commerce and the private companies Llyod Triestino and Generali, key sites for the formal materialization of urban social capital). Farnesin's network analysis does not say much, however, about the countless informal practices that rendered this cohesion possible and that allowed for the functioning of the various (formal and informal) coalitions organizing the Triestine economy and the city's politics. Yet the 'activation' of Triestine social capital has always passed exactly through a series of informal practices, tied

Figure 7.3 Galleria Tergesteo, a historical site for the Triestine urban elite (photo by Claudio Minca)

to specific (informal) geographies and languages; a unique (urban) 'habitus' showing an extraordinary resilience through the years.

The Birth of an Urban Elite

Trieste's nineteenth-century urban elite operated in an atmosphere of widespread consensus and enjoyed a significant degree of stability. Nonetheless, it never lost its will to innovate and to incorporate ever new elements, as Millo (1989) notes in her work on the formation of the Triestine bourgeoisie. The success of Trieste's new bourgeois elite came both from its ability to structure itself *internally* through a series of shared values and horizontal ties, and its capacity to legitimize itself *externally*, thanks to its facility in encompassing the entire city in its narrative of a ('new') *triestinità*. The model of civic and cultural development promoted by this new elite soon comes to coincide with the *triestinità* itself, gathering within it (albeit in diverse forms and expressions) all of the city's inhabitants and their diverse identities and self-representations. In this context, the role of the state is limited to the laying down of laws and constraints, while it is the dialect and the *triestinità* that determine the everyday practices and the informal 'rules of engagement' that order economic activity and that assure continuity and reliability (in other words, the conditions of 'trust') (Millo 2001).

Paradoxically enough, it is the city's diverse ethno-religious communities that act as the guarantors of the values that order social relations in the Triestine emporium and that, precisely for this reason, are able to dialogue with other networks of relations, external to the communities themselves: 'by appealing to the values and behaviours that the various members of the ethno-religious communities already held dear prior to their arrival in Trieste, [the different communities] allow for the formation of networks of communication and exchange based precisely upon reciprocity and trust in those very values, giving birth to new webs of economic relations as well' (Millo 2001, 388). Family ties and ethno-religious belonging thus become the building blocks of a broader urban economic community that, eventually, absorbs them both. The distinguishing characteristics of the 'Triestine' urban actor, typified by a specific ethos, a specific set of rules of professional behaviour and a specific economic culture, emerge precisely within the shared space of the *triestinità*, where the diverse actors and urban communities experiment new forms of coexistence (Millo 2001). Ethno-religious 'nation' and social class thus come to represent the key elements that discipline urban society – and that provide the idioms within which this latter organizes and represents its interests (Andreozzi 2003, 192). According to Andreozzi (2003, 195), it is exactly its capacity of determining the social conditions for the appropriation of (urban) resources and providing a framework for cooperation between diverse groups and individuals that consolidates the predominance of the urban bourgeoisie:

> its capacity to produce a context into which the new immigrants could inscribe their daily lives, [a context] within which they could find protection and the necessary ties and alliances, [a context] within which they too could represent their interests and envision future careers, but also – and perhaps in Trieste especially – a context within which merchants and entrepreneurs could prosper, a context based on unspoken guarantees and networks of trust. (Andreozzi 2003, 195)

In this context, the dialect represents, on the one hand, a necessary instrument of integration and, on the other, the language that best represents the civic and cultural values of the *triestinità* – values that coincide precisely with those of the new urban elite. The process of cultural integration that occurs through the adoption of the *triestinità* is indeed quite unique, reflecting the specificity of Triestine capitalism: 'municipal' in its idiom and in its claims to (urban) autonomy, and yet highly internationalized in its self-representations and economic horizons (Andreozzi 2003). The reproduction of Trieste's cosmopolitan bourgeoisie and its practices of integration can be conceived, as Millo (2001) argues, as a didactic process, introducing new-comers to specific social and economic practices and 'educating' them within a common patrimony of 'empirical' knowledge about 'the ways in which things are done', thus allowing them to construct for themselves not only a specific professional identity but also to gain social authority and respect and, at the same time, build much needed personal and professional networks. It is such education, both formal and informal (thus also the acquisition of the dialect) that determines the membership – and recruitment – rules of the Triestine business community: it acts as a powerful selector of human resources but also allows for a significant degree of social mobility (through, for example, the creation of various professional schools).

Unified by a particular brand of secular solidarity, the plurinational Triestine society – the cosmopolitan *Trieste Nazione* – is bound by a series of strictly delimited norms of behaviour and collaboration, bringing together the city's diverse social classes (Andreozzi 2003, 142). The *triestino* dialect is its habitus; the *triestinità* its field. The distinct urban ideology of the *triestinità* and the specific nature of its shared language and culture allow, indeed, for collaboration, cooptation and contamination between the city's different social classes or, better yet, allow for the emergence of one distinct class that claims to represent and defend the 'fundamental' values (*patria* and *famiglia*) and aspirations of all Triestines. This hegemonic urban vision (still present today in some of the representations that the city projects of itself) becomes an enormously powerful social glue that translates the *triestinità* into a discursive formation that colonizes practically all cultural expressions of Triestine urban life, from the elite to the most popular and banal (Andreozzi 2003). If we understand citizenship as the distinct point of view of the dominant that constitutes itself as universal – the point of view of those who dominate the city and who have constituted their point of view as natural by representing the city as a unity (as Engin Isin argues, 2000, 275) – the principle of citizenship that becomes dominant in Trieste corresponds to a distinct set of values and everyday practices embodied in the dialect. The *triestinità* thus imposes itself as that which Isin (1999) would define as an 'image of the city as a unified order', an order that acquires an ontological superiority over its protean, fragmented, and fractious realities. Isin suggests, indeed, that such unified representations are 'neither a falsehood nor an intentional misrepresentation. Rather, it is a solidaristic and agonistic strategy that enabled dominant segments of dominant groups to constitute themselves as virtuous, righteous and moral agents, justified in inculcating their vision in other groups' (1999, 281). In Trieste, this process corresponds to the emergence of the idea of the *triestinità*, an idea and ideal within which every enactment of a dominant Triestine social capital will be grounded from that moment on, constituting a veritable 'bourgeois' monopoly of the imagination (and, accordingly, of the possibilities for action).

We can thus see how in nineteenth-century Trieste cultural and economic capital becomes a key factor in the production of (urban) social capital and of the *triestinità* itself. The use of the dialect for everyday, informal practices, a familiarity with (usually several) European languages to be adopted in external, formal dealings, partaking in the various manifestations of Triestine civic life, are all necessary requisites for full participation in the economic and cultural life of the city – and for 'being Triestine'. The dialect becomes, in fact, an incredibly powerful tool in the bourgeois urban project which colonizes entirely Trieste's popular imaginary, forging the city's culture and self-representation in its very image. The dialect also becomes the embodiment of the Triestine *cosmos*, of its determination to remain a European, Adriatic, Mediterranean city that rejects the territorial logics of the various nationalisms that knock at its doors and that soon will transform the European political space forever (see Vivante 1912).

The Decline of the 'Cultural' Nations and the Emergence of a New Triestinità

Trieste is a city above all: a bourgeois city of international trade and finance that refutes the ties of territory and nation. The advocates of the *Trieste Nazione* speak *triestino* because they conceive of culture not as the essence of a people, as *Kultur*, but as practice, as the everyday negotiation of meaning within a much broader common project that does not require all-encompassing, definitive belonging. The dialect is, in many ways, a fluid, light-hearted, mode of communication, allowing the *Triestini* to conceive of politics and culture with that pinch of irony and disbelief that necessarily characterizes all cosmopolitan experience. The nineteenth century Triestine urban subject, deliberately without a past, without roots, speaks her/his language of origin at home – but most often chooses to experience the city in *triestino*. The *triestino* thus becomes the space that both unites (the *Triestini*) and divides (from the rest); that demarcates everything that does not 'fit' into a specific urban project, into a specific urban culture. This magical equilibrium will be shattered, however, with the triumph of nationalisms, as the *triestino* dialect comes to be seen as simply a 'local' version of a hegemonic national language, i.e. Italian, first in the liberal-nationalist imaginations of the late 1800s and early 1900s, and subsequently in the official narratives of both Fascist and, later, post-war Italy (see Apih 1966; 1988; Ballinger 1999; Hametz 2005; Matiussi 2002; Schriffer 1996).

The role played by the dialect in those decades appears, at first glance, paradoxical. What is curious, indeed, is that all through the nineteenth century the constant influx of newcomers did not seem to pose a threat to the 'Italianness' of the city: if anything, it appeared to consolidate it. The paradox lies in the fact that Trieste became an important Italian city precisely thanks to its cosmopolitanism, thanks to its civic vision and business ethic that allowed newcomers to, quite literally, work towards a common goal, to become a single community (Pellegrini 2001, 294–5). Ethnic identities were seen as something largely located in the private sphere of sentiments, and belonging in no way presupposed communities bound by common ethnic descent. Still at the outset of the 1800s, it was often difficult to clearly define the 'nationality' of many members of the Triestine merchant elite (Verginella 2001, 443), and it was only in the second half of the nineteenth century that political nationalism would gain dominance as the only possible container of ideals, values and belonging. As Verginella remarks, with the 'nationalization' of Triestine politics, both horizontal social ties, as well as the vertical bonds determined by family relations, are substituted by the ties of nationhood. The Triestine bourgeoisie feels its Italianness threatened both by increasing centralization on the part of the Empire, but also the national awakening of its Slavic neighbour (a 'threat' compounded by the emergence of a Slovene bourgeoisie within the city itself – see Schiffrer 1946; Verginella 2001). The *Triestini* thus decide to throw in their lot with fellow Italians living under the Austrian yoke, as nationalism becomes a new means of legitimizing – and promoting – the interests of the urban elite. This shift is most evident after 1867, as the political structure of the Empire undergoes a fundamental transformation (see the discussion in Negrelli 1978, 181). The idea (and ideal) of the Trieste Nazione is progressively abandoned in favour of the idea of Italianness, adopted by Trieste's liberal elite as a new hegemonic representation of – and for – the city. The myth of Triestine 'municipalism' and the very idea of *triestinità* are accordingly re-inscribed within a national(ist) logic (Valdevit 2004).

The urban elite expresses its interests in much the same ways still during the second half of the nineteenth century, yet it does so in a political context that has undergone a profound transformation:

> in this [new] context, the adoption of the idiom of nationality and nationhood on the part of the local elite appears, on the one hand, to be largely instrumental to the defence of Triestine autonomy from increasing incursions on the part of the Imperial State and the protection of local economic interests. On the other, it becomes a way

> to regain a leading role in governing the emergence of a new working class, and of a new sort of Triestine citizen. In fact, the choice of [Italian] nationality was open also to foreigners, and it was considered a cultural choice. (Andreozzi 2003, 210)

In the last decades of the 1800s, cultural 'nation' and social class – the erstwhile pillars of the Triestine cosmopolis – are thus transformed to fit the new historical-political conditions. They are translated into new coagulations of class and nationhood, though in large part still bound to preceding Triestine myths, now re-interpreted through a 'political' national lens (Andreozzi 2003). These decades witness, in fact, numerous attempts at educating the urban masses, promoting an (Italian) model of urban living capable of capturing the allegiance of the population and assimilating it to the interests of the dominant group; projects often promoted in the name of defending Trieste's Italianness. These attempts are sustained by a new network of organizations and associations that engage a large part of Triestine society in the crafting of a new urban vision, a new *triestinità*. The groups that emerge in those years encompass educational and cultural organizations, sports clubs, and a variety of leisure-time associations, whose names speak for themselves: Pro Patria, Lega Nazionale, Società di Minerva, Società Ginnastica, Università Popolare (ibid.). Many of these organizations exist still today, maintaining the traditions of Triestine associational life; all are places where the idiom of choice is still, of course, the *triestino*.

It is in these new forms of association that a new version of the *triestinità* is born, one more tighly bound to an Italian national(ist) vision, and begins to penetrate all of urban society, creating a presumed bond between the national-liberal elite and the city's population. Sport and other associational activities thus come to play an important role in the new unitary ideology of the Adriatic port. It is important to note, moreover, that this new variant of Triestine irredentism now manifests itself not only in anti-Austrian but also anti-Slavic terms, and produces as sort of permanent mobilization of Triestine society which, from that moment on, will fully embrace a series of protective measures in the name of defending its 'national spirit' (Valdevit 2004). This explains, in part at least, the triumph of territorial(ized) nationalisms from the 1860–1870s onwards, and their dominance of Triestine politics from that moment on; a triumph that will allow Fascist ideologies to take firm root in the city (see Matiussi 2002), transforming it for ever, and relegating the Triestine cosmopolitan experiment to a folkloric, bygone era.

The dialect does not disappear, however, in the tumultuous decades that mark the turn of the century, and nor do the ideas and markers of identity that gave it birth, although these are now largely consigned to the sphere of private relations. As Finzi (2001a) argues, while the triumph of the 'historical nations' may overtly undermine the role of the dialect, this legacy persists in countless everyday practices, and in the variety of informal ways in which one enters into the *triestinità*. Spaini is of a similar opinion (1963; 294), suggesting that even during the most tragic and bloody years of the twentieth century, 'the dialect continues to hold strong. It is enough to look at the popular songs of the turn of the century: far from mere entertainment, these are powerful instruments of persuasion, that reverberate in the *osterie*, burst into everyday life, and implant themselves indelibly into the memories [of the *Triestini*]'.

Triestine Irredentism

At the outset of the 1900s, the politics of the city's autonomist forces is transformed, once and for all, into a radical irredentism, and the *triestinità* becomes simply an accessory to nationalist posturing, at least in its formal manifestations. The causes of this shift are multiple and cannot be interrogated in sufficient depth here (for an analysis see Rumpler 2003). Certainly, Vienna's changing policies after 1860 play an important role in provoking an increasingly hostile stance on the part of the Trieste City Council; also crucial are the new identificatory claims of the city's Slovenian community, that add to the qualms evoked by the 'national awakening' of the Slovene nation across the border (see Schiffrer 1946; Valussi 1973; Cattaruzza 2007). What is most important for our understanding of the genealogy of the relationship between the *triestinità*, the dialect and social capital, however, is the fact that by the eve of World War I, irredentist claims come to be irrevocably wedded to the Triestine collective imaginary and become an important form of cohesion among the city's diverse social classes (see Vivante 1912). The nationalist imaginary triumphs in definitive fashion over cosmopolitan Triestine culture. Or so it seems, for while the unique form of municipal cultural irredentism that previously demarcated the city's identity and its cosmopolitan ideals is progressively marginalized, it does not disappear entirely (see Karlsen and Spadaro 2006). It is simply relegated to the private, banal spaces of the dialect and the *triestinità*, silenced by the triumph of nationalist ideologies that will become fully institutionalized with the end of the war and the Italian conquest of Trieste and, subsequently, with the rise of the Fascist regime.

The nationalist irredentism of the interwar period profoundly reconfigures Trieste's urban elite, while the ideology of the Fascist state, thanks to its grasp on the imaginations of the lower classes, robs the city's bourgeoisie of its absolute monopoly on the representation of Trieste's Italianness (see Apih 1966; Apollonio 2004; Sluga 1994; 2001). Nonetheless, the 'hidden' cultural space of the *triestinità* continues to be embodied in the erstwhile heirs of the city's nineteenth century cultural irredentism, important figures such as the Giani Stuparich (2003), Scipio Slataper (1968) and Carlo Michelstaedter (1982) and, even more generally, in the work of a whole community of writers and intellectuals who, despite their unquestioned belonging to the Italian national tradition, see in the *triestinità* a unique cultural space able to reconcile Latin, Slavic and German influences, a space kept alive all through the preceding decades precisely thanks to the Triestine *lingua franca*. It is perhaps unsurprising then that during his long stay in Trieste even James Joyce is said to have enjoyed speaking in dialect. Umberto Saba (together with Italo Svevo, one of the two leading Triestine authors of the twentieth century) also chooses to make his most famous character, Ernesto, speak in dialect. As both Saba and his later critics would claim, it is the ideal means of expression for the banal and everyday, a real, substantial language, the language of practice, of daily life (Cattaruzza 2005).

This understanding of the role of the dialect in the 'dark ages' of nationalism is best expressed in the work of Claudio Magris, the preeminent analyst of the nature of the *triestinità* that, with the collapse of the Habsburg Empire becomes a sort of 'early retirement' from the world and its worries, a permanent state of mourning for the passing of a bourgeois cosmopolitan dream that saw in Trieste its ideal laboratory. As Magris argues in the book he co-authored with Angelo Ara, *Trieste: Un'identità di frontiera* (1982), it is those writers who managed to live the city's heterogeneity to the full, who managed to capture its multiplicity in unity 'that best understood that Trieste – as the Habsburg Empire, of which it was part – was actually a model of the heterogeneous and contradictory nature of European modernity itself'. Italo Svevo and Umberto Saba both saw in Trieste, in fact, 'an ideal barometer of the intellectual and spiritual tumult about to sweep the European world; a bourgeois city par excellence, whose very history was indelibly tied to the rise – and decadence – of its bourgeoisie' (1982, 6).

This tragic sense of mourning, Ara and Magris suggest, transforms the places of the Triestine literary imagination – the city's stunning cafes, the Karst plateau, the sea – into mythical spaces of a Trieste 'that is a promise of happiness that never comes' (1982, 78–9). It is the consciousness of a forever unreachable horizon that still today profoundly marks Triestine culture, characterized by a sense of irony and a pervasive attitude of scepticism and detachment from the 'political' and all other 'serious' matters. The unitary bourgeois project that was the foundation of the myth of the cosmopolitan Trieste Nazione recomposes itself only in myth, in stories – and in the dialect.

> The neurosis of the Triestine intellectual derives from [...] the conflict between the collective desire – on the part of the city's intelligencia – for the 'Italian option' and the loss of Triestine cosmopolitanism that the Fascist interpretation of the 'national option' brings. (Ara and Magris 1982, 134)

It is the dialect, then, that becomes the 'hidden space' within which the cosmopolitan urban myth and the 'true' *triestinità* take refuge, and within which they survive during the twentieth century; no longer the privileged tool of social integration, the dialect is relegated to the intimate spaces of private relations, and the privately held perception of that which was, but that perhaps will never return.

> The only totality, the only unity that remains, with the refusal of the official, 'national' one, is the Oedipic, subterranean womb of the dialect, spoken by all social classes, adopted for all purposes, from the most banal everyday conversation to lofty intellectual discussion. (Ibid., 33)

The *triestinità* thus finds its only conduit, its only tie with a common past and an imagined common future, in literature that embodies the crisis of the city – and in the everyday use of the dialect. It is a sense of belonging that is confirmed in practice, daily, through the universal adoption of the local idiom that is so thoroughly taken for granted that still today, many *Triestini* (even highly educated ones) are unable to express certain ideas, certain feelings in 'the language of books and maps' that they were forced to learn in school.

One example that can be cited is the phenomenon of the *Lista per Trieste*, a unique political movement that has not received much scholarly attention, at least in Italy (see Cecovini 1968). The Lista was an autonomist political movement, born in the 1970s as a reaction to the signing of the Treaty of Osimo (1976) that definitely consigned the Istrian peninsula and a large part of the territories of the Venezia Giulia to the Yugoslav state, and that also called for the construction of a trans-boundary free-trade industrial zone. The movement's leaders succeeded in gathering over 60,000 signatures opposing the industrial project – and when faced with the complete disinterest on the part of the national parties, decided to form a political party to run in the next elections. The Lista swept the 1978 national elections, claiming first place and challenging the supremacy of the Democrazia Cristiana that had dominated local politics since the end of World War II.

The Lista per Trieste is, arguably, the first Italian case of an urban political movement. But its success should only be surprising to those unfamiliar with the city's historical past. According to Valdevit (2004), the Lista represented a strange re-enactment of the national-liberal coalition that had long dominated the politics of the Adriatic city, envisioned first in 'municipalist' and anti-Austrian terms and, subsequently, anti-Slavic ones. Although anti-Slavic rhetoric was not altogether absent from the Lista's rhetoric, above all it invoked the city to gather around its symbols (the 'Melon' and the 'Halberd'), to rise against national politics, against Rome and the centralizing Italian state, in a political imaginary that served up a uniquely Triestine mix of populism, anti-politics, anti-capitalism and environmentalism (Valdevit 2004). The Lista's politics (and political objectives) were, in fact, of a completely different persuasion that those of the Lega Nord, the most successful of Northern Italy's autonomist movements to emerge in the decades that follow (see Bialasiewicz 2006).

With the emergence of the Lista, Triestine social capital took on an explicitly political form, gathering together to oppose a project that, in the perception of many citizens (even those who do not end up voting for the Lista), went against the city and its 'natural vocation'. I do not have the space here to comment extensively on the political fortunes of the Lista (see http://www.listapertrieste.

it): what is important to note, however, is that as soon as the protest died down (having accomplished it political objectives), the 'Lista effect' dissolved, and the *triestinità* found refuge, once again, in the informal, banal relations of the everyday. Another example can be identified in the political career of noted coffee entrepreneur Riccardo Illy, twice the mayor of the city (elected in 1993 and 1997), and subsequently the regional governor (from 2004 to 2008). Illy, also someone from outside the national party lines, gathered broad-ranging support that cut across traditional party political and ideological divides, confirming (as some analysts have argued) the strength of Triestine popular feeling that, in determinate historical and political circumstances, is able to galvanize the city to speak with one voice, demonstrating an extraordinary capacity of rediscovering – and asserting – this supposed 'togetherness'.[3]

Figure 7.4 The Molo Audace, a key site of the informal enactment of Triestine social capital (photo by Claudio Minca)

Conclusions

In Trieste, as the *Triestini* will often point out to 'outsiders', everyone speaks entirely too much of Trieste. Whether in good terms or in bad, the city is the

3 For a critique of such understandings of the city's contemporary identity, see Valdevit (2004).

centre of conversations, always and everywhere. The Trieste immortalized by poets and novelists, a city of unresolved tension and unaccomplished desire, is reflected in ironic and banal fashion in everyday banter, in countless everyday exchanges where anything can be explained (and often is) by recourse to the presumed 'character' of the city, a character usually heavily criticized by the *Triestini*. Umberto Saba once described Trieste as a 'beautiful woman that married a rich banker;' this beautiful woman is now a widow, trapped in her memories of a time that was or could have been (Ballinger 2003b). And yet the memory of such a glorious past – in many ways maintained through the continuity of the dialect and its imagined spaces – is lived with a sense of detachment in its everyday materializations, with an ironic and disenchanted view of the city's past history and its future destiny. The omnipresent Triestine jokes and songs (Comuzzi 2003) perhaps best embody the intersection of formal and informal 'memory-work' in celebrating the Triestine *koine*: while on the one hand they recall and re-enact a profound sense of nostalgia for an idyllic past that never was, on the other, they materialize such nostalgia in the present, rendering it a moment of shared sociability.

This explains, at least in part, why the activation of Triestine social capital, now as in the past, passes through a series of informal practices materialized within a determinate series of real and imagined places. My emphasis on the role of the dialect in structuring the functioning of Triestine social capital highlights the fact that the activation of the networks of trust that render the immaterial dimension so crucial in the realization of any economic or political project is based, above all, in a series of informal practices enacted *in place*. Triestine social capital thus not only expresses itself in a series of spatial practices but activates itself, 're-awakens' (as though it were an always existing possibility, an always existing potential), through a specific language which, in that particular moment and place, makes possible a 'connection' between the individuals, the city and its memory, allowing for the elaboration of future projects (see Bialasiewicz forthcoming; Favretto 2004: Illy 2006).

It is here that the hegemonic power of the dialect still lies – but also its slippery, ambiguous nature. For while the dialect is marked by a very clear set of politics of inclusion and exclusion, as a vernacular, unwritten tongue, it is also, necessarily, a space of constant (re)negotiation and (re)formulation, which relies on endless recontextualizations and which 'makes sense' only in certain places and moments. In other words, the *triestino* is, more than anything else, an on-going performance of power and social cohesion, an 'open' space of social intercourse (for those admitted to its ranks, of course) that feeds on a vast legacy of 'things already said and done;' a space, nonetheless, that cannot be fully determined, that relies upon ever new interpretations and inputs, that must be incessantly re-activated *in place and time*. For most *Triestini* to speak in Italian – described in dialect as 'parlare in lingua', 'speaking in the language' – signifies engaging with a specific, national *space*, while speaking in *triestino* evokes *place* or, better yet, the countless real and imaginary places of their *triestinità*. It is also in this sense that we can understand the dialect as representing the *habitus* of a specific form of social capital that is tightly bound both to a distinct historical imagination but also to countless everyday manifestations in the present day. As such, it is eternally ambiguous, unspecifiable, impossible to represent – apart from the brief moments in which it is explicitly enacted in order to 'activate' social capital.

Interpreting social capital through the *triestino* dialect has two advantages (although these could also be seen as its analytical limits): on the one hand, as I have tried to suggest, in the remote past as well as today, the *triestino* has been both an elite *and* a popular idiom. The *triestino* has always been perceived – and popularly (re)presented – as the language of *all* of Trieste and, indeed, continues to represent the sort of hegemonic, unifying force for which it was 'invented' during Imperial times. In this sense, the *triestinità* (conceived as a shared way of thinking, working, being) finds in the dialect both the vehicle and the container that grants it cohesion, identity and a common communicative space within which to engage the city and its culture as a whole. At the same time, the dialect, as a sort of implicit discursive formation, by its very nature tends to co-opt and normalize the city's diverse cultural components and its diverse ways of expressing sociability. In other words, the very fact that if one does not speak dialect some functions and some roles are significantly more difficult to accomplish (if at all possible), highlights the extraordinary force – but also the limits – of a social capital that is so tightly bound, so tightly structured by its linguistic logic.

The second interesting aspect of this unique linguistic milieu derives from the fact that most *Triestini* are essentially bilingual (and, in the case of the Triestine Slovenes, trilingual), for despite the close resemblance between the *triestino* and official Italian, the negotiation of the passage between these two communicative spaces structures all social relations: *triestino* is spoken with family members, Italian is the language of documents; *triestino* is spoken with close friends, Italian with those from 'outside' (Trieste); *triestino* is the language of laughter and sorrow, Italian that of political debate and schooling; you sing in *triestino*, but read in Italian. This linguistic duality marks every aspect of daily life in the city, determining the cultural and political context of every encounter, of every relation – indeed, of the nature of communication itself, incessantly moving, as it does, between the 'formal' and the 'informal', sometimes even in the space of a single conversation.

The key point that I would like to raise here is that it is impossible to understand the functioning of this distinct form of (Triestine) social capital without engaging with the dynamics that structure the interaction between formal and informal communicative space and that act to activate social capital in specific situations and with respect to specific projects. It is precisely within an unwritten, uncodified idiom – in everyday practice expressed through particular body language such as hand gestures, through the choice of clothing or posture, all in some way evoking a certain *triestinità* – that this sort of social capital finds its measure of success.

In conclusion, let us return, then, to where we began: that is, the idea that dialect should be envisioned as a territorialized form of 'habitus', key to the production of the *triestinità*, the 'field' that allows for the activation of social and cultural capital. As Bourdieu has argued, the main purpose of the concept of habitus was to 'break away from the myth of the rational agent' (Bourdieu and Wacquant 1992). Rather, as Isin and Wood suggest (1999, 40), his theory of practice 'construes practice as a product of practical sense, of a socially constituted sense of the game'. Social capital and the networks of trust associated with it are, indeed, always activated through/in practices. And practices can never be fully described or explained. This is, perhaps, the most important conclusion – but also limitation – of research of this sort. The dialect is, as we have seen here, a unique communicative horizon. It is with respect to this horizon that agents wishing to activate social capital position themselves, as does the context that renders it effective: in other words, the ensemble of conditions that must be made to come together in order to create possibilities of collaboration and creativity in Trieste.

What is more, the *triestino*, as a language of practice and of place, as a language that is not reducible into text, is never fully mappable – it always remains (em)placed, in the sense given to this term by Doreen Massey (1993). It is interesting to note in this regard the paradox expressed by an urban culture that still today, two and a half centuries later, finds in the *triestinità* its preferred way of engaging with the world and its challenges, while the nationalist ideologies that played such an important part in marking the city's recent past and its 'official' narratives appear as the most detached, most abstract form of representation of the Triestine daily existence. For while the city's inhabitants may have been viscerally anti-Slavic in their politics and official pronouncements, all through the post-war years they continued to cross the Iron Curtain (that lay just a few kilometres from the city centre) almost daily, whether to buy cheap petrol or meat, to go out to dinner with friends, or go to the seaside, over-riding – in practice – the 'hard' cartographies of the Cold War and of nationalist ideologies. For over 40 years, the *Triestini* have continued to do business (and even marry) 'the Slavs' across the border – the sworn enemy and the 'radical other' in official rhetoric.[4]

It is useful to return to Bourdieu's work here and, in particular, an aspect of his work recalled by Isin and Wood (1999, 40):

> The cases of discrepancy between habitus and its specific inertia, its hysteresis, into the picture, are what make the analysis of habitus important. This mutual conditioning or intersubjectivity takes place in a field, which is a meaningfull

4 For a reflection on the political and cultural implication of this trans-border relationships see Bufon (1993; 1996); for an empirical analysis see Battisti (1979).

Figure 7.5 Ponterosso, a strategic landscape for the production of Triestine urban culture (photo by Claudio Minca)

> world, a world endowed with a sense and a value, in which it is worth investing one's capital.

It is precisely this almost ephemeral dimension of social capital – torn between practices and their representations – that I have tried to highlight here, all the while aware of all the limits of such a partial analysis.

There are, in fact, two key limits to a reflection on the role of the *triestino* in the formation of social capital: firstly, the fact that to understand and study this phenomenon one must speak *triestino*, which renders it practically impossible to distinguish the examining subject from the practicing subject (as, I trust, my description has made clear); secondly, the very fact that oral practices cannot be described in textual language, indeed, that they are essentially untranslatable into conventional academic terms, within which this analysis inevitably is structured.

Nonetheless, such limits are implicit in every research endeavour that aims to interrogate that which Glennie and Thrift (2005) term 'communities of practice'. Every production/activation of social capital, always and everywhere, is suspended within this relentless (and unmappable) play between the formal and the informal realms, between 'official' language and dialect, between space and place. To what extent my analysis of the activation of social capital can be applied to other contexts, marked as it is by Trieste's very distinct history and the unique role of the *triestinità*, is an open question. However, I do hope that it serves to highlight the power of informal practices and of the everyday oral performances that make possible the activation of networks of trust; by ignoring these, one risks producing a superficial account that disregards the very substance of a phenomenon that is key to the production of (material and immaterial) wealth and social cohesion.

Bibliography

Agnelli, A. (2005), *La genesi dell'idea di Mitteleuropa* (Trieste: MGS Press).

Andreozzi, D. (2003), 'L'organizzazione degli interessi a Trieste (1719–1914)', in R. Finzi et al. (eds).

Apih, E. (1966), *Italia, fascismo e antifascismo nella Venezia Giulia* (Bari: Laterza).

Apih, E. (1988), *Trieste. La storia politica e sociale* (Bari: Laterza).

Apollonio, A. (2004), *Venezia Giulia e Fascismo* (Gorizia: Libreria Editrice Goriziana).

Ara, A. and Magris, C. (1982), *Trieste: un'identità di frontiera* (Turin: Einaudi).

Ballinger, P. (1999), 'The politics of the past. Redefining insecurity along the "world's most open border"', in Weldes et al. (eds).

Ballinger, P. (2003a), *History in Exile: Memory and Identity at the Border of the Balkans* (Princeton, NJ: Princeton University Press).

Ballinger, P. (2003b), 'Imperial nostalgia: Mythologizing Habsburg Trieste', *Journal of Modern Italian Studies* 8:1, 84–101.

Ballinger, P. (2004), 'Exhumed histories: Trieste and the politics of (exclusive) victimhood', *Journal of Southern Europe and the Balkans* 6:2, 145–59.

Battisti, G. (1979), *Una regione per Trieste* (Udine: Del Bianco).

Bialasiewicz, L. (2006), 'The geographies of production and the contexts of politics: Dislocation and the new ecologies of fera in the Veneto *città diffusa*', *Environment and Planning D: Society and Space* 24:1, 46–67.

Bialasiewicz, L. (forthcoming), 'Europe as/at the border: Trieste and the meaning of Europe', *Social and Cultural Geography*.

Bourdieu, P. (1990), *The Logic of Practice* (Stanford, CA: Stanford University Press).

Bourdieu, P. and Wacquant, L (1992), *An Invitation to Reflexive Sociology* (Chicago: University of Chicago Press).

Bufon, M. (1993), 'Cultural and social dimensions of borderlands. The case of the Italo-Slovene trans-border area', *Geojournal* 30, 235–40.

Bufon, M. (1996), 'Social integration in the Italo-Slovene Gorizia transborder region', *Tijdschrift voor economische en sociale geografie* 27, 247–58.

Camuzzi, D. (2003), *La Triestinità* (Trieste: Lint).

Catalan, T. (2001), 'Presenza economica e sociale degli ebrei nella Trieste asburgica tra Settecento e primo Novecento', in R. Finzi and G. Paniek (eds).
Cattaruzza, M. (1989), 'I conflitti nazionali a Trieste nell'ambito delle questioni nazionali nell'impero asburgico 1850–1914', *Quaderni Giuliani di Storia* 10:1, 131–49.
Cattaruzza, M. (1995), *Trieste nell'Ottocento: le trasformazioni di una società civile* (Udine: Del Bianco).
Cattaruzza, M. (2007), *L'Italia e il confine orientale* (Bologna: il Mulino).
Cecovini, M. (1968), *Del patriottismo di Trieste. Discorso di un triestino agli italiani nel cinquantesimo della redenzione* (Milan: All'insegna del pesce d'oro).
Cervani, G. (2006), 'Intorno al cosmopolitismo triestino: le memorie di Giovanni Guglielmo Sartorio', in P. Karlsen and S. Spadaro (eds).
Comuzzi, M. (2003), *La Triestinità* (Trieste: LINT).
Covacich, M. (2006), *Trieste sottosopra* (Rome: Laterza).
Cusin, F. (1930), *Appunti alla storia di Trieste* (Trieste: Tipografia Triestina).
De Mauro, T. (1963), *Storia linguistica dell'Italia Unita* (Laterza: Bari).
Dubin, L. (1999), *The Port Jews of Habsburg Trieste: Absolutist Politics and Enlightenment Culture* (Stanford, CA: Stanford University Press).
Faber, E. (2003), 'Territorio e amministrazione', in R. Finzi et al. (eds).
Favretto, I. (2004), 'Italy, EU enlargement and the "reinvention" of Europe between historical memories and present representations', *Journal of Southern Europe and the Balkans* 6:2, 161–81.
Ferrari, L. (2001), 'La presenza economica degli ordini religiosi a Trieste', in R. Finzi and G. Paniek (eds).
Finzi, R. (2001a), 'La base materiale dell'italofonia a Trieste', in R. Finzi and G. Paniek (eds).
Finzi, R. (2001b), 'Trieste perchè', in R. Finzi and G. Paniek (eds).
Finzi, R. (2003), 'Introduzione', in R. Finzi et al. (eds).
Finzi, R, and Panjek, G. (2001), *Storia economica e sociale di Trieste, Vol I: La città dei gruppi 1719–1918* (Trieste: Lint).
Finzi, R., Panariti, L. and Panjek, G. (2003), *Storia economica e sociale di Trieste, Vol II: La città dei traffici 1719–1918* (Trieste: Lint).
Fogar, G. (2005), 'Il laboratorio della catastrofe: 1919–1940', in M. Rossi (ed.).
Foschiatti, M. (2006), 'Il problema nazionale della Venezia Giulia', in P. Karlsen and S. Spadaro (eds).
Gatti, C. (2001), 'Uomini e politiche nella Trieste del Settecento', in R. Finzi and G. Paniek (eds).
Geoffrey, C. (1977), *The Race for Trieste* (London: William Kimber).
Glennie, P. and Thrift, N. (2005), 'Revolutions in the times. Clocks and the temporal structures of everyday life', in D. Livingstone and C. Withers (eds).

Hametz, M. (2005), *Making Trieste Italian, 1918–1954* (Woodbridge: Boydell Press, Royal Historical Society).
Hertmans, S. (2001), *Intercities* (London: Reaktion Books).
Illy, R. (2005), 'Euroregione: il futuro a portata di mano', in M. Rossi (ed.).
Isin, E. (2002), *Being Political: Genealogies of Citizenship* (Minneapolis: University of Minnesota Press).
Isin, E. and Wood, P. (1999), *Citizenship and Identity* (London: Sage).
Jonsson, S. (2000), *Subject Without Nation* (Durham, NC: Duke University Press).
Kacin-Wohinz, M. and Pirjevec, J. (1998), *Storia degli sloveni in Italia, 1866–1998* (Venice: Marsilio).
Kaiser, A. (ed.) (1999), *La bildung ebraico-tedesca del Novecento* (Milan: Bompiani).
Kaplan, R. (2000), 'Conflict and compromise among borderlines identities in northern Italy', *Tijdschrift voor economische en sociale geografie* 91, 44–60.
Karlsen, P. and Spadaro, S. (eds) (2006), *L'altra questione di Trieste* (Gorizia: Libreria Editrice Goriziana).
Kent, N. (2007), *Trieste: a Twentieth Century Tragedy* (London: Hurst).
Livingstone, D. and Withers, C. (eds), *Geography and Revolution* (Chicago: University of Chicago Press).
Magris, C. (1996), *Il mito asburgico* (Turin: Einaudi).
Magris, C. (1997), *Microcosmi* (Milan: Garzanti).
Magris, C. (2007), 'Noi e il mondo nuovo', *Il Piccolo*, 27 December.
Massey, D. (1993), 'Power-geometry and a progressive sense of place', in J. Bird et al. (eds).
Matiussi, D. (2002), *Il partito nazionale fascista a Trieste* (Trieste: IRSML).
Merku, P. (2001), 'La presenza slovena nella città preemporiale', in R. Finzi and G. Paniek (eds).
Michelstaedter, C. (1982), *La persuasione e la rettorica* (Milan: Adelphi).
Millo, A. (1989), *L'Elite del potere a Trieste: una biografia collettiva 1891–1938* (Milan: Franco Angeli).
Millo, A. (2001), 'La formazione delle elite dirigenti', in R. Finzi and G. Paniek (eds).
Millo, A. (2003), 'Il capitalismo triestino e l'impero', in R. Finzi et al. (eds).
Minca, C. (forthcoming) 'Trieste nazione and its geographies of absence', *Social and Cultural Geography*.
Morandini, G. (ed.) (1989), *Da te lontano: cultura triestina tra '700 e '900* (Trieste: Dedolibri).
Negrelli, G. (1978), *Al di qua del mito. Diritto storico e difesa nazionale nell'autonomismo della Trieste asburgica* (Udine: Del Bianco).
Novak, B. (1970), *Trieste, 1941–1954: The Ethnic, Political and Ideological Struggle* (Chicago: University of Chicago Press).

Pellegrini, R. (2001), 'Per un profilo linguistico', in R. Finzi and G. Paniek (eds).

Pupo, R. (1999), *Guerra e dopoguerra al confine orientale d'Italia (1938–56)* (Udine: Del Bianco).

Purvis, M. and Waley, P. (forthcoming), 'Trieste: Between late-lasting Empire and late-developing nation state', *Social and Cultural Geography*.

Putnam R. (1993), *Making Democracy Work: Civic Traditions in Modern Italy* (Princeton, NJ: Princeton University Press).

Putnam, R. (2000), *Bowling Alone: the Collapse and Revival of American Community* (New York: Simon & Schuster).

Rossetti, D. (1815), *Meditazione storico analitica sulle franchigie della città e Porto Franco di Trieste* (Venice).

Rossi, M. (ed.) (2005), *Istria riscoperta* (Rome: Ediesse).

Rumpler, H. (2003), 'Economia e potere politico. Il ruolo di Trieste nella politica di sviluppo di Vienna', in R. Finzi et al. (eds).

Saba U. (1953), *Ernesto* (Turin: Einaudi).

Sapelli, G. (1990), *Trieste Italiana. Mito e destino economico* (Milan: Angeli).

Schiffrer, C. (1946), *Historic Glance at the Relations between Italians and Slavs in the Venezia Giulia* (Trieste: Stabilimento Tipografico Nazionale).

Schiffrer, C. (1992), *Dopo il ritorno di Trieste all'Italia* (Udine: Del Bianco).

Schiffrer, C. (1996), *Antifascista a Trieste* (Udine: Del Bianco).

Sestan, E. (1998), *Venezia Giulia. Lineamenti di una storia etnica e culturale* (Udine: Del Bianco).

Slataper, S. (1968), *Il mio Carso* (Milan: Maier)

Sluga, G. (1994), 'Trieste: ethnicity and the cold war, 1945–54', *Journal of Contemporary History* 29, 285–303.

Sluga, G. (2001), *The Problem of Trieste and the Italo-Yugoslav Border: Difference, Identity, and Sovereignty in Twentieth-Century Europe* (Albany, NY: State University of New York Press).

Spaini A. (1963), *Autoritratto Triestino* (Roma: Edizioni di Storia e Letteratura).

Stuparich, G. (2003), *Un anno di Scuola* (Rome: Il Ramo d'Oro).

Tamaro, A. (1974), *Storia di Trieste* (Trieste: Lint).

Urry, J. (2000), *Sociology beyond Society* (London: Routledge).

Valdevit, G. (2004), *Trieste. Storia di una periferia insicura* (Milan: Mondadori).

Valussi, G. (1972), *Il confine nordorientale d'Italia* (Trieste: Lint).

Verginella, M. (2001), 'Sloveni a Trieste tra Sette e Ottocento. Da comunità etnica a minoranza nazionale', in R. Finzi and G. Paniek (eds).

Vivante, A. (1912), *Irredentismo adriatico* (Trieste: Italo Svevo).

Wenger, E. (1999), *Communities of Practice* (Cambridge: Cambridge University Press).

Chapter 8
Boundaries of Trust: Building a Transnational Space in Haparanda-Tornio

Jouni Häkli

Introduction

The Tornio River, running through what was once the northernmost part of the Kingdom of Sweden, is a waterway whose meaning changed dramatically in 1809 when Sweden lost its 'eastern provinces' to Russia in the Finnish War. Up to that point the river had been a 410-km-long inland watercourse that provided means for mobility and a livelihood for the predominantly Finnish speaking inhabitants of the Tornio River valley. In 1809 the river gained a new function as an international boundary, first between Sweden and the Russian Empire and then, after 1917, between Sweden and the independent Republic of Finland (Lundén and Zalamans 2001).

In the delta where the Tornio River meets the Baltic Sea, a town was founded in 1621 by Gustav II Adolf of Sweden and named after the river. The location was an age-old marketplace where trade routes met and goods from the Northern wilderness, mainly salmon and fur, were transported and exchanged. The town was officially founded on the island of Suensaari in the Tornio River delta, but in the 1809 demarcation of the state boundary the Czar Alexander I specifically demanded that Tornio be annexed to the Russian Grand Dutchy of Finland. Consequently, the Swedish side of the river was left without a population centre. To compensate for the loss, Sweden started to build a new town (originally named Carl-Johan stad) by the Tornio River and in 1821 Haaparanta village (present day Haparanda) was established as a market place on the Swedish side. The village received town charter in 1842 and thus the gradual development of the twin towns of Haparanda and Tornio began (see Figure 5.1, p. 122).

Northern trade started to decline in the eighteenth century as the state boundary increasingly cut Tornio off from the position of trade hub, and the town descended from a lively merchant town into a border settlement with substantial military presence by the czar's army. However, the bridges connecting the western and eastern riverbanks were never demolished, and during World War I Tornio and Haparanda experienced a short-lived renaissance as a border crossing point for goods and people. By this time the Tornio River had already gained all the three distinctive socio-cultural and material dimensions that characterize it today: a dividing boundary, a linking meeting point, and a 'more-than-human' actor as a watercourse (Whatmore 2006).

Finland gained independence from Russia in 1917, and the country's subsequent modernization, industrialization and general urbanization pushed Tornio into further decline despite steady population growth. The bulk of industrial growth related to wood processing concentrated in Kemi, a town located at the mouth of the Kemi river only 25 km southwest from Tornio. Similar relative marginalization occurred on the Swedish side of the Tornio River where Haparanda was struggling for economic survival in the face of growing depopulation of Northern Sweden.

The border formalities between Sweden and Finland became increasingly relaxed after World War II, a process that culminated in 1996 when both countries became part of the European 'Schengenlandia'.[1] Under these circumstances, both challenging and enabling, it is understandable that the towns of Haparanda and Tornio have sought to join forces in their regional and local development efforts. It is such development efforts, deriving from local history and utilizing physical, social and cultural resources, that this chapter sets out to explore.

Attempts to explain governmental and economic development with social and cultural reasons have a long history in social thought. Authors such as Alexis de Tocqueville (1835) and Max Weber (1930) are widely regarded as forerunners in the study of the cultural roots of Western political and economic history. Their mode of thinking has encouraged generations of scholars to study the role of cultural contexts and social life in the functioning of political and governmental institutions. A recent upsurge of this research occurred in the late 1980s and early 1990s in the context of the concept of social capital (Coleman 1988; Putnam 1993; see Cannone in this volume).

Given the prominence of Putnam's (2000) idea of 'bridging' in social capital research, and the deep resonance of the metaphor of 'bridge' in border research (Sidaway 2004; Häkli 2004), it is surprising how little scholarship exists on transnational networks of trust (for exception, see Tilly 2007). In some research on international cross-border cooperation the significance of trust is noted but not analysed in depth (Church and Reid 1999; Krätke 1999; Grix and Knowles 2002). Trust is typically seen in broad terms as a factor that influences the willingness of actors to engage in institutional or semi-institutional forms of co-operation. Much less attention has been paid to the role of trust as conditioning social interactions across international boundaries. It seems almost as if the

1 The area covered by the Schengen agreement is a group of countries between which there is no security check given to people or goods crossing the borders between them.

connection between (dis)trust and (the lack of) interaction across national boundaries was too obvious to merit explicit analysis.

This chapter explores the question of transnational trust in the context of an extraordinary and challenging project that aims at physically uniting the centres of two towns located in two different countries (Finland and Sweden). With conceptual tools from actor-network theory I seek to address in a novel way the manner in which trust is related to the boundary that acts both as a force of divergence and a point of contact in international networks related to cross-border cooperation. I begin by outlining the idea of trust and then move on to discussing (international) boundaries and boundary objects as factors that may be consequential for trust-building related to successful international cooperation as exemplified by the På Gränsen–Rajalla project. I then explore the historical development of cross-border cooperation between Haparanda and Tornio at the Swedish-Finnish borderland in the southern end of the Tornio River Valley. I read this history focusing particularly on the På Gränsen–Rajalla project and seeking traces of the Tornio River as a boundary object that helps build trust among the participants of networks of cross-border cooperation as well as the broader population. To conclude, I discuss the three meanings of the Tornio River/boundary/object in light of transnational trust.

River as a Boundary as a Boundary Object

The concept of trust is routinely used in everyday language, which makes it appear deceptively simple. As a theoretical concept, however, it is complex and elusive. In social capital literature, trust is seen as a resource for social action based on its capacity for making things run smoothly because of lower 'social transaction costs' (Trigilia 2001; Veenstra 2002). In other words, where trust prevails there is less need for complicated institutional measures and procedures in order to secure the interaction between parties that ultimately are vulnerable in regard to each other. Trust is a condition that lubricates social interaction by making it easier and less mediated by pre-set institutional norms (Newton 2004).

There is now an extensive body of multidisciplinary literature discussing how best to conceptualize trust. Definitions of trust abound and there seems to be little hope for a theoretical consensus about its meaning. An extensive survey of the various uses and definitions of trust is certainly beyond the scope of this chapter (for a broader discussion on trust, see Chapter 1 in this volume). However, it is possible to limit the scope of the concept by defining it on the basis of the case in which it is to be applied. First, I am mainly interested in trust as emerging in contextual social interaction, and not as an innate psychological trait of an individual. Second, I am interested in trust as an aspect of interaction in networks between actors who have a mutual interest.

In the context of these conceptual limitations I adopt a relatively widespread working definition of trust as an actor's acceptance of vulnerability based upon positive expectations of the intentions or actions of another person, group or institution (Rousseau et al. 1998). As is evident from this definition, the possible targets of interpersonal trust range from family members to persons of a different nationality. Empirical studies on trust have shown almost unequivocally that, when asked, people report diminishing degrees of trust when social distance between the trusting person and the target of trust grows (Buchan and Croson 2004). This, of course, has clear implications for the analysis of trust in the context of international borderlands. In addition to experience gathered from concrete interaction related to cross-border cooperation, the cultural and social milieu within which this interaction takes place is pertinent to interpersonal trust. Place-bound cultural discourses concerning inclination toward trusting, such as local 'identity talk' and national stereotypes, may strongly condition actors' dispositions toward trusting, as well as reciprocating trust (Hudson 1998; Mohan and Mohan 2002; Entrikin 2003).

Theories of trust remain obscure on the issue of what gives rise to trust (Fukuyama 1995; Sztompka 1999; Veenstra 2002; Brewer et al. 2004). While the concept of a 'boundary object' certainly cannot provide a once-for-all solution, it may turn out useful because it enables both social and material realities to be included in the analysis. The dualism between nature and society has, of course, been a recurrent theme in attempts by actor-network theory scholars to decentre established understandings of what counts as 'the natural' (Callon 1986; Latour 1993). From this point of view natural entities can not be reduced to social constructs established through practices of cultural signification because they may acquire agency that is as consequential as that of social human beings (Latour 1993). Natural elements, such as rivers, may become 'agents' as part of complex and heterogeneous networks or assemblages that bring together various human and more-than-human actors, meanings, spaces, objects and materials. This agency may be unpredictable from a human point of view, and it is certainly often misrecognized as a form of agency (Kortelainen 1999).

In assessing the multiple roles of the Tornio River simultaneously as an ecosystem, as a socio-material divider, and an interface between two socio-cultural systems, I adopt the concept of boundary objects from actor-network theory (Star and Griesemer 1989). Developed as a tool for understanding how different 'communities of practice' with specific systems of meaning and ways of doing things can cooperate under circumstances of heterogeneity and diversity, the concept serves well the present analysis. It facilitates an approach where cross-border cooperation is understood as a place-bound, socio-material accomplishment that not only seeks to bridge the two sides of the boundary, but also makes use of the Tornio River as a boundary object.

Star and Greisemer (1989, 389) set forth the concept of boundary objects as an attempt to develop further the model of translations as developed by Bruno Latour, Michel Callon and John Law. They argue that the latter authors have overemphasized the perspective of the aspiring scientist or other entrepreneur who attempts to enroll allies by re-interpreting their concerns to fit their own programmatic goals (Poutiainen and Häkli in this volume). Instead of such mono-perspectival accounts of translation Star and Griesemer propose the idea of boundary objects as a means for creating coherence needed for cooperation in the face of 'very different visions stemming from the intersection of participating social worlds' (Star and Griesemer 1989, 396).

According to Star (2002) boundary object is any concrete or abstract element that people can use as a point of reference in their interactions. This sets two simultaneous demands for boundary objects. First, in order to enable common practice they have to be loosely structured and flexible so as to accommodate differences. Second, in order to retain significance and attractiveness they have to be specifiable through refinements in the context of particular locations and practices without altering the underlying consensus. Boundary objects are, thus, 'both ambiguous and clear, at different moments, for different purposes' (Star 2002, 118).

> Boundary objects are … both plastic enough to adapt to local needs and the constraints of the several parties employing them, yet robust enough to maintain a common identity across sites. They are weakly structured in common use, and become strongly structured in individual-site use. These objects may be abstract or concrete. They have different meanings in different social worlds but their structure is common enough to more than one world to make them recognizable, a means of translation. The creation and management of boundary objects is a key process in developing and maintaining coherence across intersecting social worlds. (Star and Griesemer 1989, 393)

To suggest that the Tornio River can be seen as a boundary object in the context of cross-border cooperation requires some clarification. First, what I am proposing is that the concept of boundary object may help in understanding the successes and failures of ambitious cross-border cooperation, such as the På Gränsen–Rajalla project, and moreover, in pinpointing a mechanism that fosters interpersonal trust in transnational cooperative networks. To this end I draw parallel from Star and Griesemer's (1989) study of scientific work involving the social worlds of different professionals and amateurs, thus requiring cooperation beset by the challenging heterogeneity. I am not suggesting that cross-border cooperation between two cities is similar to scientific work in regard to the establishment of natural historical facts in the context of the Museum of Vertebrate Zoology (Star and Griesemer 1989). However, there is ample evidence of the fact that efforts to establish and carry out cross-border cooperation involve heterogeneous work bringing together many different actors and viewpoints, and that this form of cooperation is equally challenged by the need to generate common understandings and manage tensions caused by the participants' diverse social worlds (Häkli and Kaplan 2002; Perkmann 2003; Kramsch and Hooper 2004).

Second, heterogeneity in cross-border cooperation is redoubled by the fact that it brings together actors from two (or more) nation-states that for decades and centuries have been socio-spatially constructed as connected, yet unconformable, socio-cultural systems separated by international boundaries. Hence, the communities of practice that face the challenge of being able to communicate and reconcile divergent meanings of the objects of cooperation include not only groups of professionals, administrators, planners, stakeholders and lay people, but also actors inhabiting two dissimilar socio-cultural systems. It is within this social landscape of (mis)trust that different boundary objects may function as intermediaries between distinct social worlds that should be able to communicate, trust, stay committed and cooperate. Arguably, like scientific work, successful cross-border cooperation depends on trust-building boundary objects that help in negotiating and resolving potentially conflicting sets of concerns that arise from the intersection of the multiple communities of practice involved.

Third, while there certainly are several items that fulfil the properties of boundary objects in cross-border cooperation between Tornio and Haparanda, the Tornio River stands out as a particularly interesting and important one. The specific role of the river is related to its (literally) central position in the social, cultural, institutional and material setting of the cross-border cooperation and particularly the På Gränsen–Rajalla project. Performing at once three interrelated but distinct functions, the Tornio River is, first of all, one of the two remaining large free-flowing rivers in the European North. This has not only prompted the inclusion of the river in the European Union *Natura 2000* programme, but it also explains the river's peculiar agency in relation to its second function as an international boundary.

The treaty of Hamina in 1809 declared that the boundary between Sweden and the Russian Grand Duchy of Finland run along a succession of the deepest points in the riverbed of the Tornio River. Seemingly stable and objective as a 'natural boundary', the riverbed is actually a rather volatile and shifting ground for a borderline. This may be due to gradual erosion or sudden changes in the riverbed caused by floods (Biger 1988, 344). Hence, the free-flowing river acquires a peculiar geopolitical agency when its alterations lead to boundary adjustments and flood protection schemes that mobilize a complex set of regulations, measurements, mapping techniques, commissioners, vessels and technologies of construction necessary for boundary demarcation and flood regulation.

The status of the Tornio River as a seemingly natural international boundary conceals the fact that it is also a socio-material accomplishment requiring considerable effort by both human and non-human actors. This work is not delimited to the periodic adjustments of the borderline, carried out at 25-year intervals by a specific Swedish-Finnish commission, but extends to myriad practices in the socio-cultural systems ongoing in the immediate vicinity of the river and beyond. This is the third function of the Tornio River. Besides being a watercourse transporting some 350 m^3 of water per second on average, as well as a 'natural boundary' marking the edge between two societies, the river is also an intersection of multiple heterogeneous actors, objects and meanings – a liminal space at the interface of two relatively distinct socio-cultural systems: the towns of Haparanda in Sweden and Tornio in Finland.

Practices of cross-border cooperation are among the most visible and organized attempts to introduce a degree of coherence and consensus to this liminal landscape with the help of boundary objects (Star 2002). In what follows I analyse the role of the Tornio River as a boundary object in the development of cross-border cooperation in the twin towns of Haparanda and Tornio, with particular emphasis on the ambitious På Gränsen–Rajalla project. I begin by outlining the progressive nationalization of the two towns into a Swedish one and Finnish one, and then move on to looking at how communities of practice have pursued cross-border cooperation so as to bridge the social worlds created through nationalizing discourses and practices. From documents and publications related to the På Gränsen–Rajalla project I trace the ways in which the Tornio River has been crafted as a boundary object, and assess its significance for trust-building as an asset in international cooperative networks.

The Tornio River at the 'Edge of Society'

International borderlands are landscapes where establishing relationships of trust is particularly challenging. The twin towns of Haparanda and Tornio, however, belong to the more amicable European borderlands characterized by a socio-cultural milieu with a long history of informal ties across the boundary. Indeed, before nation-state building in the Scandinavian North, the Tornio River Valley was a predominantly Finnish-speaking region with strong cultural and linguistic unity. This unity was put under strain in 1809 when the Tornio River became an international boundary, but nevertheless there is still some linguistic and cultural affinity across the boundary. The border has always been very open except for unusual conditions, such as war, and today a substantial share of the population of Haparanda has Finnish citizenship. Moreover, some two thirds of the population in Haparanda speak either Finnish or 'Meänkieli' (a Finnish dialect spoken on the Swedish side of the Tornio River Valley), and the inhabitants of Tornio have all learned Swedish at school (Zalamans 2002).

However, despite the long history of interaction across the Tornio River, and the existence of a distinctive Tornio River Valley identity (Tornionlaakso/Tornedal identity), people in Tornio and Haparanda also tend to identify strongly with their respective national communities (Lundén and Zalamans 2001; Jukarainen 2003). Since 1809 a growing number of institutional and popular practices of distinction have worked to divide the region into Swedish and Finnish sides, and today national stereotypes concerning life and culture on the other side are commonly resorted to when depicting each other's habits, collective character, values and the national(ized) landscapes (Prokkola 2005). The two national master narratives, Swedish and Finnish, have thus created an order that people both in Haparanda and Tornio tend to view as natural. It is this socio-cultural landscape of (mis)trust, embodied by the Tornio River as a 'natural boundary', that various practices of formal cross-border cooperation have sought to accommodate and bridge.

While informal ties first dominated cross-border activities between the citizens of Haparanda and Tornio, after the 1960s the official relationships between the two towns started to warm up considerably. This development was initiated by public sector actors who saw the possibilities for synergy in some areas of public service provision. The process also reflected the relaxation of the Swedish and Finnish states' grip on this peaceful borderland and was backed up by the rise of the ideology of 'Nordic countries' as a unifying theme in Scandinavian international relations. The membership of Sweden and Finland in the European Union since 1995 further boosted this development, with cross-border cooperation being one of the official targets of the EU cohesion policies.

Indeed, it can be noted that the Tornio River Valley has much to show when it comes to official cross-border cooperation. The municipalities on both sides have engaged in synergetic relations for several decades, and the towns of Haparanda and Tornio are cooperating on issues ranging from the common use of a swimming pool (since 1960) to the establishment of a supranational body, Provincia Bothniensis (1987), for coordinating issues of common interest across the national boundary. Forms of practical cooperation include agreements on joint waste water treatment (1971), joint language education services (1989), joint district heating system (1993), joint tourist information office (1998) and joint preschool (2002) (Provincia Bothniensis 2008). Furthermore, the Euro has become a commonly accepted currency also on the Swedish side, despite the fact that Sweden decided to stay out of the European Monetary Union (Zalamans 2002).

Haparanda and Tornio's notable achievements in international cross-border cooperation have been acknowledged in several studies (Zalamans 2002; Ioannides et al. 2006). However, the persistence of the divisive effect of the international boundary has also been noted. For example Lundén and Zalamans (2001) point out that while political and governmental actors show a great deal of willingness to cooperate across the boundary, similar enthusiasm cannot be found among business actors or large sections of the citizenry. In studies of the youth in Haparanda and Tornio, Jukarainen (2003) notes that despite the many forms of cross-border cooperation, nation-specific identities prevail. This factor can be explained by the effective use of school education as a tool of Finlandization and Swedification since the late nineteenth century. Indicative of just how persistent the boundary is as a cultural divider, the results of a recent study show that even students attending the jointly operated bilingual school in Haparanda tend to socialize only with their own nationals outside the school activites (Teppola and Mustajärvi 2006).

The nationally divisive effect of the Tornio River presents a challenge to establishing cross-border networks of trust. This has been pinpointed, for instance, in a document prepared as a proposal for the Tornio River Valley Interreg III programme.

> In all, the national boundary and disparate state legislations have hindered or at least slowed down the emergence of effective transboundary systems. ... the main obstacles to cross-border cooperation have been the following: language problems ... suspicious attitudes ... dissimilar practices of politics and government ... lack of knowledge and contacts ... national regulations for the use of resources ... Finland's membership in the European Monetary Union. (Tornionlaakson neuvosto 2008)

The problem of trust can be seen to culminate in an observation by Lundén and Zalamans (2001, 42) that, among many groups in both Haparanda and Tornio, there is a feeling that 'what is good for them is bad for us'. Along similar lines the Interreg III programme document notes that the hindrance is not so much the unwillingness to cooperate, but suspicions about 'the neighbor benefiting more' (Tornionlaakson neuvosto 2008, 4). It is within this cultural landscape of (mis)trust that the Tornio River has come to act as a boundary object with significant effects on the cooperation across the Swedish-Finnish boundary.

Trust-building through Boundary Objects: The *På Gränsen–Rajalla* Project

In discussing the role of boundary objects in scientific cooperation Star and Griesemer (1989) found that elements performing the mediating function between communities of practice were of different kinds. They distinguished between material representations (such as maps and diagrams), settings of action (museum as a common referent), and virtual orders (ideal types, standardized forms), all of which were helpful in resolving the problem of heterogeneity in cooperation between intersecting social worlds. In what follows I describe the På Gränsen–Rajalla project (Figure 8.1) as a form of cooperation characterized by the challenge of bridging social worlds set apart by social and cultural heterogeneity. I set out to trace elements that have performed the function of boundary object by settling some of the social, cultural and material complexity of the project.

Methodologically I follow the example of Star and Griesemer (1989) who based their analysis of Berkeley's Museum of Vertebrate Zoology on various documents produced in the early twentieth century and stored at the Museum. In exploring the På Gränsen–Rajalla project I focus on the documentation

Figure 8.1 The logo of På Gränsen–Rajalla project foregrounds the unity of Tornio River (© Tornion kaupunki)

related to the planning and realization of the project as it stands in the spring of 2008. While the materials to be analysed carry certain limitations, including their emphasis on the visual and verbal at the cost of direct observation, they nevertheless enable reflection concerning the means that have been utilized to facilitate communication and cooperation across intersecting social worlds at the international borderland of Haparanda and Tornio.

The På Gränsen–Rajalla project (literally meaning 'on the border project') is an effort at coordinated community planning between Haparanda and Tornio, with the aim of physically uniting the two towns by constructing a common town centre. The project idea is presented to the public as follows:

> På Gränsen–Rajalla is co-operative project of two cities located in two different countries, Haparanda (Sweden) and Tornio (Finland). In the middle of the border line we are building a center where you can find – besides commercial services – also apartments, educational possibilities, jobs, culture, and free-time options. Building started at spring 2003. Into the new centre we are building senior apartments, common police station and a shopping center across the border line. The local streets are also rebuilt to cross the border between Finland and Sweden. The new traffic circle makes the traffic easier. … Do you want to build borderless future with us? Don´t hesitate to contact us! (På Gränsen 2008)

The background of the På Gränsen–Rajalla initiative lies in the many geopolitical changes that the collapse of the Soviet Union and the European Union membership of Sweden and Finland brought to the European North in the early 1990s. Together with a growing concern for the population loss and stagnant economic development in the Haparanda-Tornio region, new possibilities for cooperation and mobility across the Swedish-Finnish boundary encouraged the towns to become active in strategic planning (Santasalo and Heusala 2002). Important momentum for the project was given by the existing forms of institutional cooperation such as the *Provincia Bothniensis*, which had shown that Haparanda and Tornio may indeed benefit from joining forces. Yet, collaboration in the area of physical town planning was unforeseen, a point emphasized in a document reporting the results of the design competition which launched the project in 1997.

> [Haparanda and Tornio] have, since several years, been engaged in extensive cooperation, but until now cooperation regarding the physical planning has not been far-reaching. … The new situation without hindering formalities has inspired to a new and more extensive cooperation regarding the physical planning. … The two municipalities together with Road Administrations in both countries, therefore decided to arrange an architectural competition. (Arkitekturtävlingar 1998, 16)

On the basis of two winning designs, a Development Plan which focused on the un-built, idle land between the existing town centres of Haparanda and Tornio was drafted in the spring of 2000. At this stage the key actors in the binational cooperation network were Seppo Pelttari (the chairman of Tornio town council), Gunnel Simu (the chairman of Haparanda municipal council), Bengt Westman (the chairman of Haparanda local government), Raimo Ronkainen (the acting mayor of Tornio), Janerik Reyier (director of the department of roads, Region Norr, Sweden), and Tapani Pöyry (director of the department of roads, Lapland, Finland) (Utvecklingsplan 2000, 1). These persons also acted as the jury of the design competition and thus were in a very strong position to set developmental guidelines for the project (Arkitekturtävlingar 1998, 2).

Within the steering group of the På Gränsen–Rajalla project, cooperation was carried out without much friction. In an article published before a consultative referendum on the project was held in Haparanda on 15 September 2002, Bengt Westman, one of the leading figures in the project network, noted that the members of the steering group had already voted 'yes' and this has brought the project both political and financial support. He also pointed out that the project had been ongoing for several years and would require further commitment for years to come (Westman 2002). Obviously the long history of interaction within the jury that later became the project's steering group had established an atmosphere of trust among the core members (Poutiainen and Häkli in this volume).

However, the problem of cross-national (mis)trust was accentuated after the planning stage when the project met the challenge of engaging new supporters from the ranks of local politicians, business communities, and most importantly, ordinary citizens in both Haparanda and Tornio. In anticipation of an unwelcome reception of the På Gränsen–Rajalla project among the Haparanda citizenry, Bengt Westman lamented that

> Unfortunately, most unfortunately, there are people who have mistaken [the project's] illustrations for descriptions of what exists instead of what they really are,

> imaginations among certain architects and consultants. Communication [concerning the real state of affairs] has also been hampered by some town council members who in wishing to show their political competence and secure their re-election have chosen to publicly portray the illustrations as reality. I utterly regret this. (Westman 2002)

Westman's fears were not groundless. Symptomatic of a broader problem of (mis)trust impeding cross-border cooperation, the referendum in Haparanda rejected the På Gränsen–Rajalla project (På Gränsen archive 2008). Perhaps fortunately for the project, no referendum was held in Tornio, and because the Haparanda referendum was only consultative, both towns decided to continue working on the project. Hence, despite the lack of popular support on the Swedish side, the master plan was accepted by both town councils in February 2003. Since then the plan has started to materialize in the actual border landscape, albeit at a varying pace (Figure 8.2).

Figure 8.2 The På Gränsen–Rajalla planning area in 2000 (© Tornion kaupunki)

Source: City of Tornio 2008.

The first concrete measures in 2003 were related to earth-moving work aiming to make the ground more suitable for building. In 2005 the project was given a boost by IKEA, a multinational furniture and accessories company, which announced its plan to open an outlet at the border between Haparanda and Tornio. The company's decision to invest in the planning area was an important step forward for the På Gränsen–Rajalla project, a project that for the critics seemed little more than a waste of public funds and lacking support among the local business actors and citizens (Asemakaava 2003). Somewhat surprisingly, the På Gränsen–Rajalla project is currently a lively development scheme that has been capable of attracting the interest of dozens of companies willing to follow the lead of IKEA (Enberg 2006).

How, then, should we understand these successful efforts to transcend the Tornio River as a national boundary by building a transnational centre that unites the towns of Haparanda and Tornio? IKEA's major investment in the area definitely took the project to a new level, but it can hardly explain how the project got through the rocky road of its early stages which were characterized by lack of support and confidence among the political and business elite, as well as the general population. To account for the success of the Steering group in enrolling new 'believers' to support this ambitious project, I suggest a close re-reading of the På Gränsen–Rajalla project documentation in order to trace the uses of the Tornio River as a trust-building boundary object. In what follows, I propose three different forms or guises in which the river has come to acquire this function: 1) the Tornio River as a *physical entity* vested with a degree of agency in relation to the human communities that inhabit its banks; 2) *representation* of the Tornio River as a means of overcoming the dividing function of the border; 3) the national boundary as a *symbolic space* in the border landscape.

This threefold typology of boundary objects follows Henri Lefebvre's (1991) distinction between spatial practice, representation of space and lived space. In referring to Lefebvre, I wish to emphasize the spatially intertwined nature of the three aspects of the Tornio River as a boundary object (Star and Griesemer 1989). Seen as a physical entity, the river is related to social and material processes that Lefebvre terms 'spatial practice', thus underlining the fact that, by necessity, all individual and social existence makes use of space. 'Representation of space' points at the useful quality of images and maps as selective and modifiable abstractions that help reduce the complexities that characterize the social, cultural and physical world. Representation of the river as a boundary may foster cooperation simply by showing how the two national sides can be brought together by means of physical planning. In Lefebvre's terminology 'lived space' means simply the spatially ubiquitous symbolism that characterizes all lived human experience. In the case of the Tornio River, this refers to the boundary as the river's affective kernel and locus of passion – as spatial symbolism that unites differing experiences in lived situations.

The Flooding River and its Counter-measures

Due to melting snow, the outflow of water in the catchment area of the Tornio River usually causes flooding that raises the river's water level by not more than 2.5 m above sea level. However, unusually warm weather and heavy spring rain coupled with ice jams in critical locations may occasionally cause flooding that exceeds the normal level. This was the case in the spring of 1990 when the water level in the Tornio River rose to the record high of 4 m above sea level, causing damage in Tornio to some buildings and streets that were covered by water (Utvecklingsplan 2000, 2).

From the point of view of the På Gränsen–Rajalla project, the flooding of the Tornio River comprises two aspects that make it a prime candidate for a boundary object. First, it is a phenomenon that occurs regularly, albeit with varying degrees of intensity, so that all actors related to the river share by necessity the consistent unpredictability of the flooding. Second, while the causes of flooding are beyond human control, its consequences can be regulated by water level surveillance and physical structures built to protect areas threatened by flood-related damage. Together these two aspects imbue the flooding river with the capacity of being simultaneously loosely structured and flexible, as well as specifiable in the context of particular practices without altering the underlying consensus. In other words, the flooding river is at once an actor and a boundary object bridging the gap between alternate social worlds that meet in the På Gränsen–Rajalla project.

The physical planning area of the På Gränsen–Rajalla project is boggy lowland between the town centres of Haparanda and Tornio. From early on, the planning has had to contend with the fact that the new joint urban centre belongs to areas most vulnerable to flooding, and thus requires special protective measures in order to be at all viable as a construction site. In commenting on one of the winning designs (shared first prize) the jury composed of the members of the core network reasoned as follows:

> The design is based on a rigorous analysis of the landscape, traffic network and existing urban structures. … The new park joining the towns … utilizes seasonally changing water levels in a controlled way by building a series of water basins that make the park unique. (Arkitekturtävlingar 1998)

Interestingly enough, while the competition assignment called for designs to construct a centre for cooperation, the jury clearly hesitated to award proposals that contained ideas of concrete urban building that would unite Haparanda and Tornio. Instead, the three winning designs all suggest that the area between the towns be developed as a park (Arkitekturtävlingar 1998). To some extent this can be accounted for by the more moderate level of ambition that first characterized the project, but arguably the threat of flooding also played an important role in raising caution among the jury and the competing architects about the possibilities for large scale construction in the area.

The situation changed with the next stage of the project when a more ambitious Development Plan was drafted, comprising the idea that the town centres be actually built together by locating some 105,000 m^2 of newly permitted building area in the vicinity. However, on top of extensive land fill needed to secure the required amount of building sites, a system of canals, basins, embankments and dams had to be designed for regulating the fluctuating water levels in the vicinity of the planning area (see Figure 8.3, which shows the arc-shaped embankment dominating the plan).

> The buildings in the [new joint] town centre will be protected from flooding by building walls and embankments. The area that is subject to flooding is [considered] from the point of view of highest water level (4.0 meters in 1990) and normal water level (2.5 meters) … To prevent water from pouring into the planning area from the Tornio River … the canals will be furnished with floodgates located at the outfall of the flues. (Utvecklingsplan 2000, 27)

While certainly more expensive than building on more solid ground, the idea of joining the two towns physically was now too precious to be subjected to economic critique. In the Development Plan, the steering group of the På Gränsen–Rajalla project states (much too optimistically, as it turned out) that

> [t]he demand of soil resources needed for land filling is estimated to reach 415.000 cubic meters. This makes possible the easy relocation of both towns' excess soil masses in the area. Most of the materials will naturally be taken from the proximate neighboring areas. (Utvecklingsplan 2000, 27)

However, before any ideas presented in the Development Plan could be realized, the plan would have to be developed into an Executive Plan accepted not only by the authorities in both countries but also, as an actor in a key position regarding the success of the project, by the Border River Commission between Sweden and Finland. The Commission is a transboundary governing body in charge of monitoring the use of the Tornio River as regulated by the border treaty between Finland and Sweden dating back to 1972. Until 2004 its jurisdiction included issues related to the building and alterations of the river basin and, hence, Haparanda and Tornio had to submit an application for permission to realize the Tornio River related parts of the På Gränsen–Rajalla project.

In a decision on 28 June 2002 the Commission granted Haparanda and Tornio the right to carry out their joint water construction project on the condition that special measures be taken to protect the building site against even the most extreme flooding, estimated to occur at the rate of no more than once in a hundred years.

> The Finnish-Swedish Border River Commission concludes that permission to the applied water construction project can be granted on the following conditions. … Flooding embankment must be built so that the water-resistant layers of the dam reach the height of 4.00 meters above normal sea level, and on top of that, the surface

> layers of the dam will sustain at least one day against a flood rising to the height of 4.35 meters above normal sea level. (Border River Commission 2002, 5)

In pinpointing the flooding and the well-planned measures to control it as decisive factors favouring the approval of the project, the Commission was not only able to communicate with the steering group on an equal plane, but it could also retain a degree of specificity in framing the issue in its own terms. Herein lies the moment of the flooding Tornio River acting as a boundary object. In bridging the gap between the governmental steering group and the legislative Commission, the flooding river reduced the socio-material complexity of the project to the level where a critical decision could be made in order to realize it.

In the Executive Plan of 2003, the Border River Commission's decision was referred to as an authoritative statement obligating Haparanda and Tornio to supervise that all water construction work fulfills the conditions set by the Commission. Again the flooding river acquires a key role in the realization of the plan because the ability to control the river's water level is the single most important prerequisite to any other building activities in the area. Hence, the Executive Plan states the following:

> In the Excecutive Plan the project has been divided into stages that will be realized incrementally over 10-20 years' time, or even faster. The first stage comprises infrastructure building that facilitates the further development of the area ... In the area embankments will be built so that excavation work can be carried out on dry land. (Toteuttamissuunnitelma 2003, 3, 20)

The actual water construction work commenced in Tornio in April of 2003 and has been ongoing since then. This may also be important when considering the interface between the På Gränsen–Rajalla project and the general citizenry. A somewhat alarming observation about the lack of popular interest in the project was made on the basis of the fact that only 17 people attended a discussion organized by the Tornio town planning department on 25 June 2002 (Asemakaava 2003, 11). The task of moving the earth as a countermeasure against flooding is where the broader citizenry have seen the first concrete traces of the project on the border landscape. Arguably the project has gained more support among the citizens simply because of the general perception of the usefulness of such public works. A growing positive attitude towards the project is evinced by the outcome of a recent local survey where nearly two thirds of the respondents reported that they believe the project to be useful (Niska 2005). Again, it is the flooding river that acts as a bridge between different social worlds. For ordinary town dwellers, as well as for the Border River Commission and the Steering Committee, the flooding river presents a common denominator that enables each social domain to frame the issue according to its own concerns while retaining a degree of consensus about the object's identity.

Cohesive Representation of Divided Space

In a statement concerning the design competition that launched the På Gränsen–Rajalla project, the jury explicitly referred to the importance of maps in representing the planning area.

> Coordinated physical planning in the borderland between Haparanda and Tornio has been practically nonexistent. Maps and charts have often ended at the border. (Arkitekturtävlingar 1998, 2)

This simple notion points to the situation that characterizes most international borderlands all over the world: the border effectively separates two nationally constructed social and political systems from each other. The division often culminates in the 'peaceful coexistence' of two distinct jurisdictions, separated by the international boundary that marks the absolute end of national authority (Häkli and Kaplan 2002). Despite extensive cross-border cooperation between Haparanda and Tornio, the Swedish-Finnish boundary makes no exception. Both towns have actively sought to establish a transnational body for the joint exercise of public powers to handle their shared projects, but as is the case with similar aspirations elsewhere, neither the Swedish nor the Finnish state has considered this as a possibility (Rajakuntayhteistyö 2002).

Overcoming the dividing function of the boundary between Sweden and Finland has not been easy in the area of physical planning. The extensive, joint planning activity notwithstanding, Haparanda and Tornio have to carry out the actual zoning as two separate processes politically regulated by two different legislations. Hence, besides the joint På Gränsen–Rajalla planning process that has been ongoing since 1997, both towns are zoning their own parts of the planning area respectively (Asemakaava 2002).

In these circumstances it is not surprising that images, maps and charts representing the transnational planning area as one unity have emerged as an important set of boundary objects that bridge the two communities of practice composed of nationally constituted social, juridical and political systems. As pointed out by Star and Griesemer (1989, 411), coincident boundaries represented on maps enable work at different sites to be 'conducted autonomously while cooperating parties share a common referent'. In this regard the maps of the På Gränsen–Rajalla planning area have been extremely important facilitators of the two systems' conjoint action.

Quite expectedly, the importance of maps and images is reflected in their large number in the existing planning materials (see På Gränsen–Rajalla 2008). Even a cautious estimate of the number of images and maps portraying the 'unity' of the planning area comes to 250, which means that, on average, each page of the available planning materials contains more than one such image. In more qualitative terms, the importance of the cohesive representation of divided space can be judged from the almost iconic nature of some of the cartographic representations depicting the area. Portrayed consistently and repeatedly throughout the documents that have been disseminated among the local population both materially and through the project's website, the map of the planning area has become a symbol in its own right, with meaning and significance that go well beyond merely instrumental (Figure 8.3). Particularly revealing is the choice by the Steering Committee to place this particular map on the front cover of the marketing brochure aimed at the business world and the general public. It is as if the transnational space, renounced by the two national governments, had been literally drawn into existence by a persistent group of enthusiasts.

Despite the transboundary nature of the envisioned space, from early on the På Gränsen–Rajalla project has sought a careful balance between the ideas of obscuring and symbolizing the national boundary. The jury of the design competition made a clear point about the need to retain a specific attitude towards the border.

> The boundary has lost its earlier meaning but nevertheless it is still a border between two states and, thus, there is a need to take a particular stance toward it. This sentiment should, at least symbolically, be visible to all those who cross the border. (Arkitekturtävlingar 1998, 15)

This liminality, a border visibly dividing what is to become a transnational space, is carefully taken into account in the design solutions of the På Gränsen–Rajalla Development Plan. Throughout the planning area the boundary line is clearly marked in the physical space by means of water channels and other indicators, even in the market square that is one of the focal points of the plan (Utvecklingsplan 2000). This already suggests the third way in which the Tornio River functions as a boundary object in the På Gränsen–Rajalla project.

A Border that Unites

In envisioning a transnational space between Haparanda and Tornio, the jury of the design competition was clearly aware of the international boundary's symbolic power and significance as a tourist attraction too. The jury pointed this out in setting up the competition task.

> In the competition programme aims are stated … 'to develop the two cities to a unity where the border does not divide but unite' … 'to gather information on communications and tourism and to host major public events'. (Arkitekturtävlingar 1998, 16)

The goals of at once dissolving and emphasizing the boundary appear contradictory, but when understood in terms of lived experience of the boundary as a boundary object, it is possible to move beyond this apparent

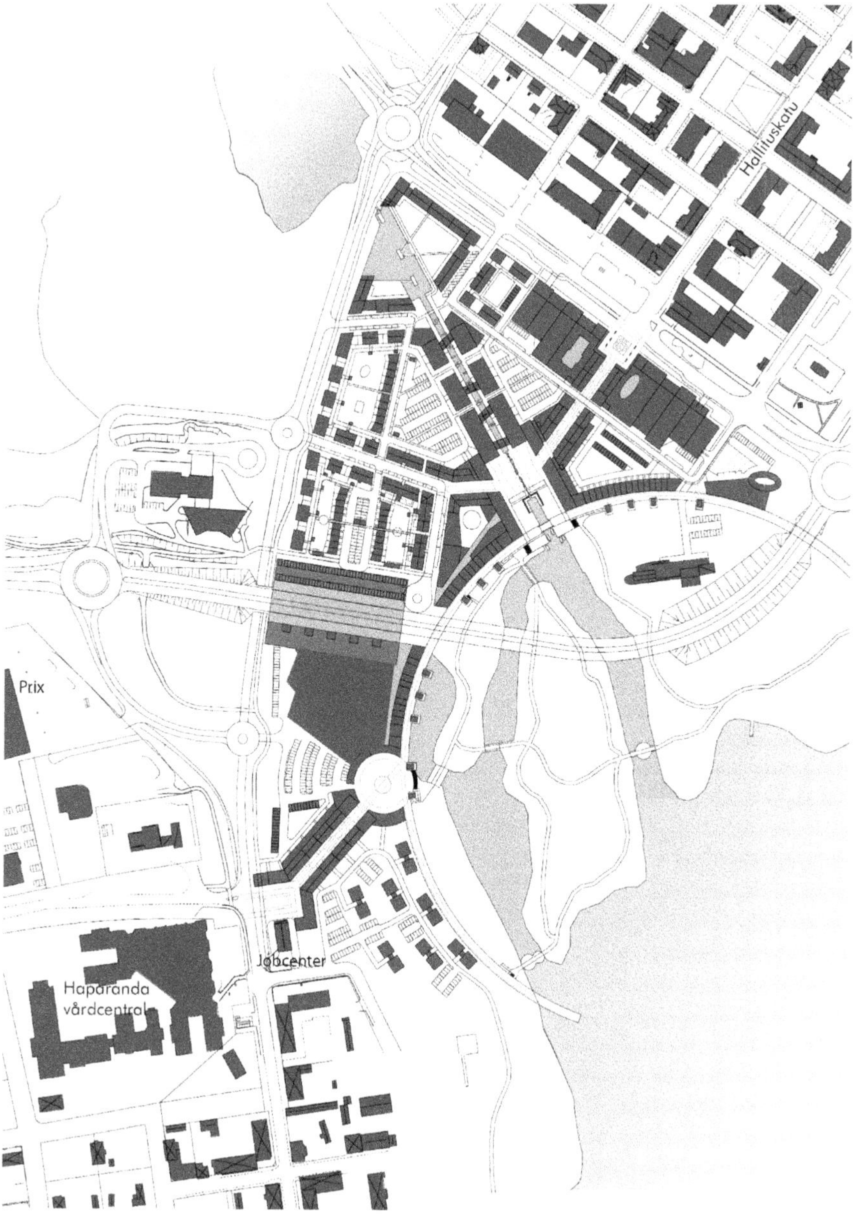

Figure 8.3 **Iconic map of the På Gränsen–Rajalla planning area (© Tornion kaupunki)**

paradox. In fact, as part of lived space, boundary acquires a symbolic function in the original Greek sense of *sýmbolon*, which means 'throwing things together', or 'contrasting'. The symbolically unifying sense of the border is explicitly underlined in the På Gränsen–Rajalla Development Plan.

> On the border a unique site for private companies, public and commercial services, knowledge center and culture will be built. The special border town will even become an important tourist attraction. The boundary will be used as a point of contact in all respects ... Even though the towns will be physically united the border will continue to be visible. The border will be highlighted in built environment, parks and market square by means of architecture, landscaping and environmental art ... differences in building styles will be observable when moving from one country to another. (Utvecklingsplan 2000, 3)

While the Tornio River cannot represent the border because of intensive building in the På Gränsen–Rajalla planning area, it is nevertheless the water from this river that marks the boundary. In this sense the steering group saw that the Development Plan would uphold the historical experience of the Tornio River as a border river (Figure 8.4). To make sure that the physical marker will actually become part of the citizens' lived experience, the Development Plan contains elements that aim at providing opportunities for ordinary people to meet and interact within the transnational space.

> Services and free time activities will be placed on the border and its immediate vicinity so that the citizens from both towns can use them, together with workplaces where contacts across the border occur daily. In the park the border is formed as a channel where it is possible to paddle, row and fish in the summer time. ... In the winter time shared skating and curling tracks will be made on the ice. ... The border remains visible but several bridges will be built everywhere to make its crossing easy. (Utvecklingsplan 2000, 7)

It is in the lived spaces of the På Gränsen–Rajalla area that, once the work is completed, the Tornio River is expected to function as a boundary object, symbolically unifying the two sides of the boundary. While the work is still ongoing, at this stage it is clear that the steering group has been able to build trust among the key actors in the På Gränsen–Rajalla project network. By enacting their social capital they have also successfully mobilized a resource that has helped overcome the social, cultural and material complexities that pertain to the production of transnational space. A question for further research to settle is whether the steering group's goal of constructing a 'border that unites' will actually be attained when the new centre eventually becomes a lived space.

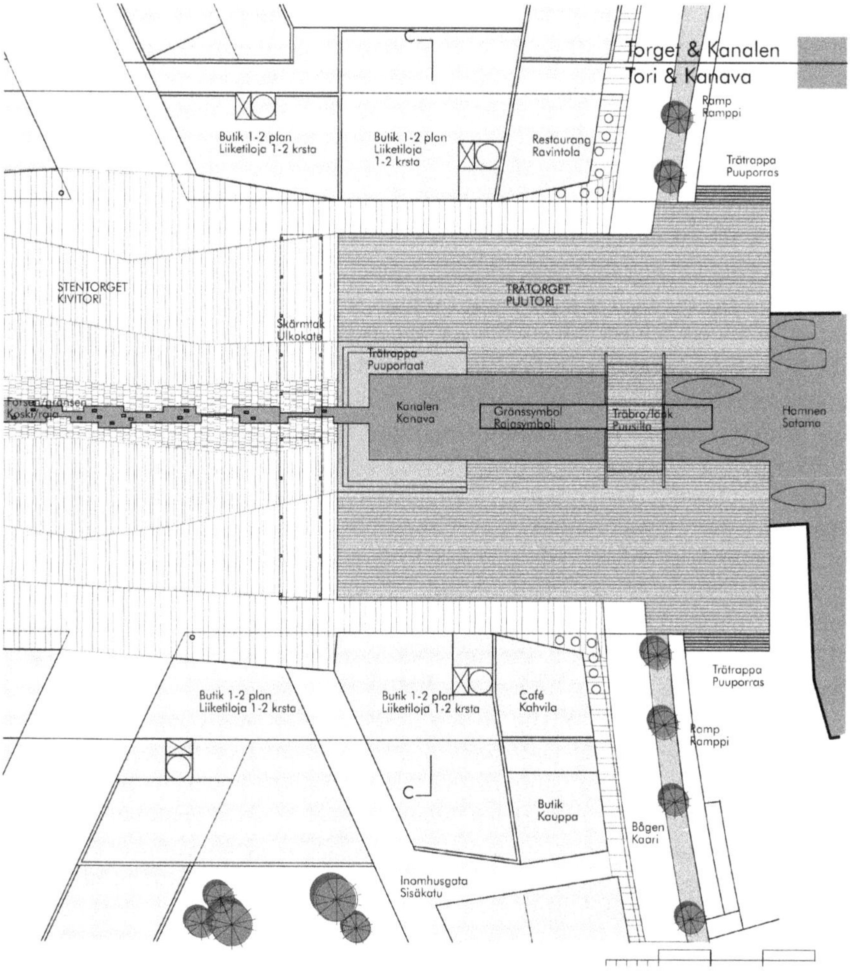

Figure 8.4 Plan for the market square with horizontal water canal marking the state boundary (© Tornion kaupunki)

Conclusion

In this chapter I have assessed the social, material and cultural complexity that characterizes a cross-border cooperation project called På Gränsen–Rajalla. I have suggested that the project can be suitably viewed by focusing on trust-building boundary objects that perform the important function of bridging communities of practice that inhabit different social and cultural domains. Despite some conceptual similarity, my assessment breaks critically with Robert Putnam's (2000, 22) thought in which a distinction is made between 'bridging' and 'bonding' forms of social capital. For him bridging refers to connections among heterogeneous people spanning 'diverse social cleavages' and constituting useful connections across social roles, interests, spaces and worldviews. The latter term he reserves for inward-looking social networks that conjoin similar kinds of people and reinforce exclusive identities. Moreover, Putnam (2000, 22) contends that bridging networks help actors to link with external assets and to diffuse information, both of which foster projects related to development and economic performance.

However, I assert that even though Putnam (2000) notes that bridging and bonding are not strict categories but rather dimensions along which different forms of social capital can be placed, the distinction is an oversimplification that is based on false separation between social relations that in reality always appear mixed (see also Leonard 2004). It thus provides inappropriate tools for analysing the complex conditions under which development projects are carried out. A case in point is the long-term effort to envision, design, plan and build a joint centre for Haparanda and Tornio as a new kind of transnational space. Whereas Putnam (2000) never explicates what he considers as evidence of bridging social capital, I have striven to show that boundary objects, existing simultaneously as social, material and cultural entities, can be pragmatically seen as trust-building mechanisms that, in the words of Star and Griesemer (1989, 414), 'are not simply the imposition of one world's vision on the rest [but rather] act as anchors or bridges, however temporary'.

In assessing the role of the Tornio River as a boundary object that helps build trust among the participants of networks of cross-border cooperation, I have stressed the original unity of 'the social' and 'the material' that in many strands of social thought have been treated separately. To this end I have sought resources from actor-network theory (Callon 1991) and Henri Lefebvre's (1991) spatial triad, both of which emphasize the complex intertwinement of social, material and cultural worlds.

In the På Gränsen–Rajalla project the representations of the Tornio River, showing its role as a boundary in images, maps and graphs, are not just abstractions but also material imprints that exist and circulate as objects among other objects in the social world. Hence, maps and images exist materially just like the objects that they represent. Conversely, the materiality of the Tornio River, as expressed for example in its yearly flooding in the springtime, is abstracted whenever the threat of flooding is taken up as an issue to be dealt with. The flood, in this case, is a discursive event just as it is a material one, occurring with some regularity. The experience of the Tornio River as a lived space, again, brings together its symbolic and material aspects, thus condensing into the experience the historical evolution of national narratives and discourses, the sense of the overwhelming material presence of the river's water and banks, and the meanings these may acquire as part of contemporary social life in all its complexity.

It is this complexity that the three types of boundary objects, each belonging to the fundamental spatiality of human existence and related to the Tornio River, have helped reduce by showing how actors inhabiting different domains can all approach what they may consider as essentially the same, but nevertheless, specifiable in ways that reflect the needs of each social world (Star and Griesemer 1989). If successful, the new centre of the twin-town Haparanda-Tornio will one day emerge as a transnational landscape – a symbolic space that people from both towns can identify with and feel comfortably part of. In such a case, people may experience a sense of belonging in their living environments instead of merely consuming the representations, master narratives and gleaming images through which the landscape is depicted in official discourses.

If international trust is likely to prevail in places where people do *not* feel alienated from the landscapes they live in (Olwig 2005), the critical question is whether transnational trust can be successfully built from the top down, as is still much the case with the På Gränsen–Rajalla project. I have proposed that the Tornio River has successfully fostered mutual trust as a threefold boundary object. However, in the case of the Haparanda-Tornio twin-town it appears that, while personal trust may be on a high level between key actors of the steering group of the På Gränsen–Rajalla project, the ordinary borderlanders may be living in a very different world, one marked by distrust toward the official efforts to gradually diminish the Swedish and Finnish national identities. In the case of the Tornio River Valley, this is exemplified by the fact that the plan to physically unite the towns of Tornio and Haparanda was rejected in a referendum in Haparanda but still pushed forward by both town councils.

This, in fact, is not unusual at European borderlands. Instead of a passionate will to exploit the possibilities of the increasingly open EU internal boundaries, many ordinary borderlanders remain indifferent toward the market opportunities 'on the other side' (van der Velde and van Houtum 2004). Moreover, there typically exists a gap between institutional cross-border activities and the borderlanders' interests toward and knowledge about these activities (Häkli 2001; Kramsch 2002; Sidaway 2004; Strüver 2004). The discrepancy between official and popular views may have the consequence of diminishing the significance of the personal transboundary trust among the professionals of cross-border cooperation. Thus, whatever the achievements of cooperation at the official level are, these may not reverberate much in the broader social sphere, leaving the achieved transnational landscape hanging in the air and without deeper rooting in the borderlanders' social and political fabric. This in itself may hinder the emergence of transnational trust in Haparanda-Tornio.

Is the På Gränsen–Rajalla project, then, likely to fail? This is not necessarily the case. IKEA's surprising decision to invest in the Haparanda-Tornio area may turn out to be one of the most important occurrences capable of bringing the project to fruition. IKEA has great potential in this respect, largely because its image fits well in the transnational town centre, and by attracting new investments in the area, it has greatly increased confidence in the project. Moreover, IKEA's products build concrete links between homes as people's most intimate spaces and the emerging transnational landscape. It is well known that IKEA deliberated upon locating its new store and eventually decided in favour of Tornio-Haparanda instead of more southern locations in Tampere and Turku. Playing along with this northern preference has certainly helped the company in winning over peoples' hearts and it may well be that IKEA eventually acquires a decisive role as a boundary object in the transnationalization of Haparanda-Tornio.

Bibliography

Arkitekturtävlingar (1998), *SAR's documentation nr 10*, <http://www.haparanda.se> (home page), accessed 15 October 2007.

Asemakaava (2003), 'Asemakaavamuutoksen selostus, joka koskee 13.11.2002 päivättyä asemakaavakarttaa' [Summary of changes to town plan 13.11.2002], *På Gränsen–Rajalla project documentation*, <http://www.pagransen.com/fin/ finstart.html>, accessed 16 May 2008.

Biger, G. (1988), 'Physical geography and law: The case of international river boundaries', *GeoJournal* 17:3, 341–7.

Brewer, P. R., Gross, K., Aday, S. and Willnat, L. (2004), 'International trust and public opinion about world affairs', *American Journal of Political Science* 48:1, 93–109.

Buchan, N. and Croson, R. (2004), 'The boundaries of trust: own and others' actions in the US and China', *Journal of Economic Behavior and Organization* 55:4, 485–504.

Bucken-Knapp, G. and Schack, M. (eds) (2002), *Borders Matter: Transboundary Regions in Contemporary Europe* (Aabenraa: IFG).

Callon, M. (1986), 'Some elements of the sociology of translations: domestication of the scallops and fishermen of St Brieuc Bay', in J. Law (ed.).

Callon, M. (1991), 'Techno-economic networks and irreversibility', in J. Law (ed.).

Church, A. and Reid, P. (1999), 'Cross-border co-operation, institutionalization and political space across the English Channel', *Regional Studies* 33:7, 643–55.

Coleman, J. (1988), 'Social capital in the creation of human capital', *American Journal of Sociology* 94, 95–121.

Enberg, J. (2006), *IKEAs etablering i Haparanda* [IKEA's settling in Haparanda], unpublished report, Södertörns Högskola <www.diva-portal.org/diva/getDocument?urn_nbn_se_sh_diva-619-1__fulltext.pdf>, accessed 5 May 2008.

Entrikin, N. (2003), 'Placing trust', *Ethics, Place and Environment* 6:3, 259–71.

Fukuyama, F. (1995), *Trust: The Social Virtues and the Creation of Prosperity* (New York: The Free Press).

Grix, J. and Knowles, V. (2002), 'The Euroregion as a social capital maximizer: The German-Polish Euroregion pro-Europa Viadrina', *Regional and Federal Studies* 12:4, 154–76.
Häkli, J. (2001), 'The politics of belonging: Complexities of identity in the Catalan borderlands', *Geografiska Annaler B: Human Geography* 83:3, 5–13.
Häkli, J. (2002), 'Transboundary networking in Catalonia', in D. Kaplan and J. Häkli (eds).
Häkli, J. and Kaplan, D. (2002), 'Learning from Europe? Borderlands in social and geographical context', in D. Kaplan, and J. Häkli (eds).
Hudson, A. (1998), 'Placing trust, trusting place: On the social construction of offshore financial centres', *Political Geography* 17:8, 915–37.
Ioannides, D., Nielsen, P. and Billing, P. (2006), 'Transboundary collaboration in tourism: The case of the Bothnian Arc', *Tourism Geographies* 8: 2, 122–42.
Jukarainen, P. (2003), 'Definitely not yet the end of nations: Northern borderlands youth in defence of national identities', *Young* 11:3, 217–34.
Kaplan, D. and Häkli, J. (eds) (2002), *Boundaries and Place: European Borderlands in Geographical Context* (Lanham, MD: Rowman and Littlefield).
Kortelainen, J. (1999), 'The river as an actor-network: the Finnish forest industry utilization of lake and river systems', *Geoforum* 30:3, 235–47.
Kramsch, O. (2002), 'Reimagining the scalar topologies of cross-border governance: Euroregions in the post-colonial present', *Space and Polity* 6:2, 169–96.
Kramsch, O. and Hooper, B. (eds) (2004), *Cross-Border Governance in the European Union* (London: Routledge).
Krätke, S. (1999), 'Regional integration or fragmentation? The German-Polish border region in a New Europe', *Regional Studies* 33:7, 631–41.
Latour, B. (1993), *We Have Never Been Modern* (New York: Harvester Wheatsheaf).
Law, J. (ed.) (1986), *Power, Action and Belief. A New Sociology of Knowledge?* (London: Routledge).
Law, J. (ed.) (1991), *A Sociology of Monsters* (London: Routledge).
Lefebvre, H. (1991), *The Production of Space* (Oxford: Blackwell).
Leonard, M. (2004), 'Bonding and bridging social capital: Reflections from Belfast', *Sociology* 38:5, 927–44.
Lundén, T. and Zalamans, D. (2001), 'Local co-operation, ethnic diversity and state territoriality: The case of Haparanda and Tornio on the Sweden-Finland border', *GeoJournal* 54:1, 33–42.
Mohan, G. and Mohan, J. (2002), 'Placing social capital', *Progress in Human Geography* 26:2, 191–210.
Newton, K. (2004), 'Social trust: individual and cross-national approaches', *Portuguese Journal of Social Science* 3:1, 15–35.

Niska, M. (2005), *Gränsen: Ungdomarnas attityder till att bo in gränsland* [Border: Young people's attitudes towards living in a borderland], Luleå tekniska universitet, Institutionen för Arbetsvetenskap 2005:25, <http://epubl.luth.se/1402-1781/2005/25/LTU-CDUPP-0525-SE.pdf>, accessed 11 June 2008.

Olwig, K. (2005), 'Representation and alienation in the political land-scape', *Cultural Geographies* 12:1, 19–40.

På Gränsen–Rajalla (2008), *På Gränsen–Rajalla* [website], <http://www.pagransen.com>, accessed 29 March 2008.

På Gränsen archive (2008), *På Gränsen–Rajalla dokumenttiarkisto* [website], <http://www.pagransen.com/dokark-dokarkisto/dokarkisto.htm>, accessed 10 April 2008.

Perkmann, M. (2003), 'Policy entrepreneurs, multilevel governance and policy networks in the European polity: The case of the EUREGIO', *Department of Sociology, Lancaster University* [website], <http://www.comp.lancs.ac.uk/sociology/soc104mp.html>, accessed 14 August 2005.

Prokkola, E.-K. (2005), 'Tornionlaakso arjen paikkana – kertomuksia Suomen ja Ruotsin rajalta', *Terra* 117:3, 177–88.

Provincia Bothniensis (2008), *Provincia Bothniensis* [website], <http://provincia.haparanda.se/>, accessed 10 February 2008.

Putnam R. (1993), *Making Democracy Work: Civic Traditions in Modern Italy* (Princeton, NJ: Princeton University Press).

Putnam, R. (2000), *Bowling Alone: The Collapse and Revival of American Community* (New York: Simon & Schuster).

Rajakuntayhteistyö (2002), *Rajakuntayhteistyön syventäminen. Työryhmän raportti* [Deepening of cross-border between municipalities. Working group report], Sisäasiainministeriön kuntaosaston julkaisu 4, <http://www.poliisi.fi/intermin/biblio.nsf/4D34CCA1E00CA5B6C2256C67005455FB/$file/rajakuntayhteistyo.pdf>, accessed 5 May 2008.

Rousseau, D., Sitkin, S., Burt, R. and Camerer, C. (1998), 'Not so different after all: A cross-discipline view of trust', *Academy of Management Review* 23:1, 393–404.

Santasalo, T. and Heusala, H. (2002), 'Haaparannan ja Tornion keskustojen kaupalliset palvelut' [Commercial services in the town centres of Haparanda and Tornio], *På Gränsen–Rajalla project documentation*, <http://www.pagransen.com/fin/ finstart.html>, accessed 16 May 2008.

Sidaway, J. (2004), 'The choreographies of European integration: negotiating trans-frontier cooperation in Iberia', in O. Kramsch and B. Hooper (eds).

Star, S.L. and Griesemer, J.R. (1989), 'Institutional ecology, "translations" and boundary objects: Amateurs and professionals in Berkeley's Museum of Vertebrate Zoology, 1907-39', *Social Studies of Science* 19:3, 387–420.

Strüver, A. (2004), '"We are only allowed to re-act, not to act": Eurocrats' strategies and borderlanders' tactics in a Dutch-German cross-border region', in O. Kramsch and B. Hooper (eds).

Sztompka, P. (1999), *Trust: A Sociological Theory* (Cambridge: Cambridge University Press).
Teppola, H. and Mustajärvi, H. (2006), *Två länder, en gräns, en skola: En studie på Språkskolan i Haparanda* [Two countries, one border, one school: A study of Languageschool in Haparanda], Luleå tekniska universitet, Institutionen för Arbetsvetenskap 2006:14, <http://epubl.ltu.se/1402-1781/2006/14/LTU-CDUPP-0614-SE.pdf>, accessed 12 March 2008.
Tilly, C. (2007), 'Trust networks in transnational migration', *Sociological Forum* 22:1, 3–24.
Tocqueville, A. de (1835/2000), *Democracy in America*. Translated by George Lawrence (New York: Perennial Classics).
Toteuttamissuunnitelma (2003), 'På Gränsen – Rajalla executive plan', <http://www.haparanda.se> (home page), accessed 15 October 2007.
Tornionlaakson neuvosto (2008), *Interreg III programme proposal*, <www.tornedalen.org/pdf_filer/Interreg%20III_fi.pdf>, accessed 20 January 2008.
Trigilia, C. (2001), 'Social capital and local development', *European Journal of Social Theory* 4:4, 427–42.
Utvecklingsplan (2000), *På Gränsen–Rajalla development plan*, <http://www.haparanda.se> (home page), accessed 15 October 2007.
van der Velde, M. and van Houtum, H. (2004), 'De-politicizing labour market indifference and immobility in the European Union', in O. Kramsch and B. Hooper (eds).
Veenstra, G. (2002), 'Explicating social capital: trust and participation in the civil space', *Canadian Journal of Sociology* 27:4, 547–73.
Weber, M. (1930), *The Protestant Ethic and the Spirit of Capitalism*. Translated by Talcott Parsons (New York: Scribner).
Westman, B. (2002), 'Vill diskutera sakfrågor om På Gränsen' [Want to talk about facts of På Gränsen], <http://eurocitynet.nu/lasmer.asp?artikelnummer=1144>, accessed 5 May 2008.
Whatmore, S. (2006), 'Materialist returns: Practising cultural geography in and for a more-than-human world', *Cultural Geographies* 13:4, 600–609.
World Bank (2005), *World Development Indicators 2005* (Washington, DC: World Bank)
Zalamans, D. (2002), 'Transboundary Regionalization – the Case of Haparanda and Tornio', in Bucken-Knapp, G. and Schack, M. (eds).

Chapter 9

All that is Solid does not Melt into Air: (Re-placing) Social Capital in Venice's Urban Development

Mauro Cannone

> [Venice] is still left for our beholding in the final period of her decline: a ghost upon the sands of the sea, so weak, so quiet, so bereft of all but her loveliness, that we might well doubt, as we watched her faint reflection in the mirage of the lagoon, which was the City, and which the Shadow. I would endeavour to trace the lines of this image before it be forever lost, and to record, as far as I may, the warning which seems to me to be uttered by every one of the fast-gaining waves, that beat like passing bells, against the stones of Venice.
>
> Ruskin 1853/1958, 13

Introduction: Thetis, or the 'Strange' Reappearance of Social Capital in Venice

On 25 August 2006, among the usual political news on the front page of the Italian daily *La Repubblica*, a particular title caught the attention of Venetian and Italian readers: 'Venice in 2030: a depopulated city' (Bianchin 2006). Reporting the results of a recent study commissioned by the Venice City Council, the article noted that the steady population decline of the last half-century was set to continue and would eventually lead to Venice becoming a mere theme park with no permanent residents, but only countless tourists admiring the ruins of what once was one of the world's most vibrant cities. Just a few weeks before the publication of this article the Venice in Peril Fund organized an international symposium at the Royal Geographic Society in London with a provocative title: 'Enough money has been spent saving Venice.' The symposium was preceded in the British press by several polemical headlines such as: 'If you love Venice, let her die' (Campbell-Johnston 2006), or 'Send for Disney to Save Venice' (McKie 2006). Such contentious propositions accentuated the key message of the symposium, namely, that there is no economic rationale for stubbornly prolonging the life of this old city by means of artificial, and very costly, treatments, as it will only delay her swan-song and inevitable disappearance under the lagoon waters. And thus, since even ordinary Venetians have decided to surrender to such an 'unavoidable' fate, why then spend money to restore the beautiful facades of her *Palazzi* or the creeping walls of her numberless churches? If anything, the most rational decision would be to turn Venice into a theme park run by a corporation – apparently the only solution whereby the city could support her final years of life.

The common thread linking these studies, proposals and provocative statements, is the problematic idea that the future of Venice was already inscribed in her stones, as John Ruskin had lamented many years ago (see opening quotation). In fact, it was as if this and the many other Romantic paeans that celebrated the subject of *Death in Venice* with joyful melancholy were correct. As if, after all, everybody knew that sooner or later *it* would happen, Venice would 'die' because poets, travellers, traders and visitors of every sort had declared that it would, as though Venice's legends of death, mystery, and ghosts had ordained that as a city born from the sea she was destined to return there. The notion of the Venice's inevitable demise, whether tentatively proposed as a solution to its problems, or decried as a threat, both in the past and in the present, has always seemed inseparable from the very idea of the city – a spectre hanging over it – as though it were a city with no other option but to wait for the end[1] (see Bec 2003; Grubb 1986; Isnenghi 1997).

What certain Romantic poets long ago lamented, numerous influential researchers, economists and journalists have recently acknowledged, although arguably with even more deterministic overtones. This notwithstanding, in March 2007 philosopher Massimo Cacciari, the mayor of Venice, publicly called for greater mutual cooperation between local institutional bodies. More precisely, he declared that there are 'concrete possibilities for the city to develop her own economy, industry, and productive sectors beyond tourism' (*La Nuova Venezia* 2007). During the same event, Paolo Costa, the former mayor and a renowned professor of economics, claimed in a similar fashion that the city has to develop her potential and become a leading worldwide technological hub (Bellemo 2007).

On that day, together with other business and political leaders, they celebrated the 10th anniversary of the opening of Thetis: a company located

1 According to Pes (2002; see also Zanetto et al. 2003), any academic study of Venice is confronted with the necessity of outlining its particular territorial view and definition of the city because in the twentieth century her modern industrial expansion towards the mainland (see Soriani in this volume) posed the question of a new and much larger administrative unit and as a consequence, two contrasting visions and definitions of the city followed. Despite the significant enlargement of the municipality, some define 'Venice' as only the historic island surrounded by the lagoon waters, while for others this is just the centre of a much wider metropolitan unit enclosing the island. The present study draws on the first approach for one simple but fundamental reason: the project of connecting every single part of the whole into a new metropolis, proposed at the beginning of the last century, has never been completed (see Benevolo 1996a; D'Agostino 2002; Zucconi 2002b).

in the historic area of the Arsenale and operating mainly in the sectors of environmental technologies and Intelligent Transport System (ITS) – a company that for them, as for many other Venetians, clearly illustrates that the socio-economic vitality of their city can be 'saved' regardless of statistical data and fatalistic predictions. And that the city can be saved precisely by countering the intransigent claims, made both by Romance poets and scientific studies that she will inevitably drift under the shores of her lagoon.

Developed from an initial idea set out in a dissertation carried out at the IUAV, the Venice university of architecture,[2] Thetis was first established as a centre for marine technologies. Ten years ago it was little more than a pioneering experiment, and arguably an ambitious project with little hope of success, as it aimed not only to reverse the decay of the Arsenale, and more broadly the socio-economic decline of Venice, but also to address the environmental weaknesses of her natural ecosystem. In less than a decade, Thetis has become a profitable shareholders company, which has worked on some of the most sophisticated projects in its areas of expertise: from the pollution of the Venetian lagoon, to the systems of traffic regulation for the municipalities of Venice, Rome, Beijing and New Delhi, and also the MOSE system of mobile barriers designed to protect Venice from high tides (see Paruzzolo et al. 2007).

However, as the local press largely emphasized on the occasion of its 10th anniversary, beside its role of innovative enterprise in a strategic sector for the city, Thetis today is also an exemplary model of successful cooperation between public institutional bodies and local private companies. With Thetis not only has a successful research and business consortium been created, but much more importantly, the problematic dependence of Venice on the tourist industry has been challenged, while the city's natural fragility has been proposed as a crucial competitive advantage.

More precisely, in these press accounts the successful constitution of the Center was associated with some fundamental social elements, such as cooperation, collective trust, and mutual support. These resembled very closely the aspects identified by Robert Putnam (1993) to explain the contemporary political and economic performances of northern and central Italy, areas that he described as successful geographical units where individual participation in public life acted as the precondition for the achievement of a stable and diffused prosperity. At first sight the outcomes of this initiative could be intuitively associated with Putnam's renowned argument, and therefore be envisaged as the spontaneous results of associative and cooperative 'social capital' firmly rooted in these 'civic regions'. Indeed, any enthusiastic disciple of this American political scientist could similarly suppose – even without reading the press accounts retracing the origins of this centre but 'just' following the firm conceptual relations established by Putnam (see Chapter 2 in this volume)

2 The second largest university in the city (after Ca' Foscari University) and one of the most renowned in Europe (Zucconi 2002a).

– that Thetis is one of the countless beneficial consequences of the 'trustful culture' characterizing the areas lying just across the waters of the Venetian lagoon. To what extent, however, is such supposition supported by empirical evidence and facts?

Certainly, upon a quick glance at the contemporary debates around the future of Venice, even the most 'orthodox' scholars of social capital would at least be puzzled by the fact that such deterministic linkages are nowhere to be found in the accounts of the recent history and geography of the city. In these debates Venice is rather compared to a noble elderly lady who wasted an immense inheritance, and whose life now depends on the generous, but lethal, assistance of national and international, public and private institutions, as well as on countless visitors queuing to admire her unique heritage in exchange for a small entrance fee (Campbell-Johnston 2006; see also Morand 2002). Put another way, the economy of Venice does not work as spontaneously and as effectively as it should, and the consequence is that this 'lady' is now almost unable to stand on her own feet.

What happened to the solid cause-effect relations that Robert Putnam revealed between history, geography, and economic growth? In other words, why did the glorious democratic past and favourable location of the city not act as preconditions for a prosperous present? In answering these questions, should we trust the alleged rationalism of recent analyses which, in leaving no hopes for Venice, implicitly suggest that these relationships have melted into air? Or, alternatively, should we envisage Thetis as the sign of their strange reappearance on the Venetian landscape? More importantly we may ask, in which direction is the future of the city heading?

By now, whoever grasps the contrast between these two opposite forecasts for the future of the city may well feel intellectually trapped between their deterministic grips. Venice in general – and the creation of Thetis in particular – may, then, seem to offer an ideal ground for reflection not only on the concept of social capital and on Robert Putnam's 'social capitalism' (Fine 2002) but also on the rationalism of influential studies and opinions, and their inability to include in their analyses a largely unpredictable element: human agency. By now, in fact, the more attentive scholars interested in understanding the relations between social capital and economic development might begin to suspect that there is something theoretically unsound in both the virtuous circle of civicness and wealth that Putnam sketched, and in the resolute suggestions to let Venice follow her Romantic fate. And for this same reason, they might also be convinced that it is probably time to test the very assumptions from which these conclusions on social capital and on Venice stem. In other words, they might agree that instead of being continuously beguiled by the plethora of statistics and supposedly 'scientific' methods that these studies employ, it is time to challenge their problematic perspectives and to recognize the empirical social processes with which local development is actually achieved, and concomitantly the role that social capital plays within them.

This is precisely what I have set out to do in this chapter, which will reflect on the constitution of Thetis as an opportunity to understand the effective ways in which the 'social' could become a 'capital' so solid and powerful as to subvert any deterministic prediction. A capital whose solidity, as I shall argue, was firmly cast within a socio-material depth that Putnam's studies, as well as the recent suggestions concerning the future of Venice, have utterly overlooked: the depth of the local 'territory' where specific resources are embedded and where social relations are continuously recreated by virtue of a spatial proximity, which is still crucial in imbuing these relations with *power* (see Allen 2003). Without attempting to produce a definitive and generalized account of these long-discussed issues, I will investigate the specific case of Thetis with the aim of acknowledging how 'development' can be empirically achieved and of recognizing how social capital can effectively contribute to the process.

Following and Re-placing Social Capital in Venice's Urban Development

As I have argued elsewhere in this volume (see Chapter 2), the plethora of figures that Putnam employs in his studies have built a very persuasive account of the close relationships that exist between social capital, space, and political and economic performance. In a similar fashion, when applied to the analysis of the socio-economic conditions of contemporary Venice, the alleged neutrality of these methodological tools might suggest that very little can be done to reverse the decline experienced by the city. The problem with such quantitative analyses, I will contend, is that whether purposely or not, they come to establish a static and passive perception of individuals and their agencies and therefore tend to overlook the external circumstances that affect subjective decisions and behaviours (see Bagnasco 1999).

The methodological tools I will employ in this study are fundamentally different for one basic but crucial reason: my aim is not to foretell the tendencies of urban development in Venice, but rather to investigate the causal relations by which one visible example of urban development occurred in this city. And if, as Callon (1986) observes, an empirical fact is (produced by) a process that cannot be measured but 'merely' described, the causal relations from which this fact stemmed are concealed in its very description, rather than in a passive quantification that excludes active subjects from analysis. Therefore, following the principal idea lying at the heart of the concept of social capital – an idea envisaging social relationships (the 'social') as a useful resource ('capital') to pursue a given goal – what needs to be problematized is not whether or not, in Venice, there is a sufficient 'stock' of social relationships that could make the economy work (this might indicate merely the number of relations and the probability of their usefulness), but rather the ways in which these relations are, in fact, deployed.

The key question here is *why* social relationships affect the empirical production of the (urban) economy in Venice and elsewhere. This is a problem that in my view requires, by definition, the use of qualitative sociological tools that – following Bourdieu's (1984; 1986) conceptualization – are fundamental for capturing the processes by which a wide range of resources can be accessed through social relations, as well as for framing social capital within the places in which it is (re)produced and utilized by individuals and groups working on actual instances of 'development'. With these qualitative tools I aim to explore the role of social capital in the constitution of Thetis – an initiative of urban development implemented, as I shall demonstrate, by means of a place-specific form of social capital that, in turn, depended on the 'local' social and geographical context in which this initiative took shape.

In other words, I will attempt to follow social capital in the real 'places' in which it was enacted, and this means recognizing and taking into account its peculiar social and geographical situatedness. More precisely, this means removing it from statistical data aggregated into the administrative units of regions or nations (see Chapter 2 in this volume) and 're-placing' it within what geographer Giuseppe Dematteis (1993; 1999) defines as the 'local milieu' – a distinctive and always unique social and geographical environment comprised of a set of particular qualities embedded in a territorial context (see also Conti 1993; Governa 1997; Giaccaria 1999). In this conceptualization, the natural, social, and 'cultural' qualities of the milieu are envisaged as the effective resources of local development because they provide the basis on which the opportunities of endogenous change and evolution rest. The effective translation of these potential resources into actual development depends, however, on a set of social relations established between those local actors who control such resources. I would argue that it is within this set of subjective relationships that social capital might play a fundamentally active role, because it can influence the ways in which these same actors associate in networks operating within and beyond the milieu, and therefore, it can enable or hinder their ability to mobilize the wide range of resources required to pursue developmental initiatives.

Moreover, in this conceptualization the milieu emerges not only as the 'container' where a wide set of socio-material resources are embedded, but also as the local spatial dimension where social networks are effectively assembled, that is, as a place which 'favors the constitutions of relations between actors that are socially (and spatially) close' (Conti 1993, 121). The relationship between social capital and socio-economic growth can thus be captured within the 'local milieu', because it is here that existing acquaintances, contacts, and mutual recognition between actors help the constitution of networks through which, similarly to Bourdieu's (1986, 248) definition, actual or potential resources are mobilized, managed, and put at the disposal of other network members (see Governa 1997).

But how is the 'local network' empirically created and held together by social relations, and how does its social capital enable the mobilization and translation of 'local' resources into real development? To answer to these questions, as Governa and Salone (2005, 278) have recently argued, the investigation of developmental practices requires a geographical perspective focused on the territorial dynamics that generate development, that is, on the 'social' processes by which networks of local actors revalorize the components of their milieu as unique competitive advantages. For this reason, in order to address the territorial dynamics with which Thetis was translated from an idea into concrete reality, and to frame the role that social capital played in such dynamics, I will draw on the 'slow' sociology of the actor-network theory (hereafter ANT) as the methodological basis for that 'slow' process of 'geo-graphy' that I intend to retrace here.

In the methodology of the ANT, as outlined in particular by Bruno Latour (2005) and Michel Callon (1986), the focus is placed on the empirical processes and on the visible actors of the 'social' world, because these processes and these actors are the central elements that ANT recognizes as legitimately belonging to such a world. Instead of viewing society as being enclosed and at the same time explained by an invisible 'social' dimension existing before everything, ANT treats social facts as material things whose origin and composition can be fully traceable with a 'sociology of association', i.e. a methodological approach whose main concern is to follow and (re)trace the processes that produce the momentary associations among social agents (Latour 2005). In ANT vocabulary, these processes and associations take the metaphorical shape of networks of heterogeneous actors that are the one and only protagonists of society. Indeed, the ANT, in Latour's (1996, 369) own words, 'does not wish to add social networks to social theory but to rebuild social theory out of networks' (see also Latour 1986; Law 1994).

In following these methodological suggestions and attempting to translate them within a geographical context, I will retrace the processes by which some of the social and material components of the Venetian milieu have intertwined into the networked association(s) from which Thetis was originally created. I will also provide a meticulous account of all the successive phases and passages in which these elements have interacted. It is with this perspective that I intend to analyse the empirical establishment of Thetis, making fully visible the actors, the processes, the resources and in one word, the network(s) from which such a paradigmatic example of urban development has been implemented. In this way, I intend to re-trace a 'slow' process of geo-graphy performed by a network of actors in which the social and material elements of the same milieu have established a set of dynamic associations developing, transforming, and 're-inscribing' the urban territory of Venice. Also semantically, the very idea of *transformation* implies a 'slow' transition into a new status – the process of *development* towards a successive condition – that I intend to uncover.

Following this method, I will re-trace here four crucial moments of network building that Callon (1986) identified in his seminal account on the 'sociology of translation'. I shall argue that Thetis is the result of the progressive unfolding in space and time of these moments: the *problematization* of the decay of the Arsenale of Venice; the *interdefinition of the actors* that supported the particular solution of the problem; the *device of interessement* deployed to strengthen the alliances between these actors; the *enrolment and mobilization* of further resources required to complete the project. Throughout this processual development different human actors and non-human elements, territorial resources and social capital were gradually enticed into and assembled within a heterogeneous network, which was slowly taking its shape and achieving its goals.

The Four Moments of Translation

> Every time you want to know what a non-human does, simply imagine what other humans or other non-humans would have to do, were this character not present. This imaginary substitution exactly sizes up the role, or function, of this little character. (Latour 1992, 229)

The 'Problematization' of the Arsenale

As the name suggests, the Arsenale of Venice is a large area[3] of the city where at the beginning of the XII century shipbuilding activities were gathered and put under the direct control of the Venetian Republic (Zanelli 1983). Because of its innovative methods of mass-production, the Arsenale was eventually defined as an *ante litteram* Fordist factory (Aymard 1991; Piva 2003) which in the 16th century, at the peak of its industrial activity, employed approximately 4000 workers who could assemble almost one commercial or military ship per day. For these reasons the Arsenale has always been described as the real secret of the successful expansion and domination that Venice established on the Mediterranean Sea (Aymard 1991; Concina 1991). Probably the most advanced Centre of production in Europe for many centuries, after the fall of the Republic in 1797 it slowly became a semi-abandoned area largely destroyed by the Napoleonic army and kept in a marginal economic position during the successive Austrian domination of the city. With the annexation of the Veneto region and of Venice (1866) to the unified Italian state (1861), the Arsenale regained part of its former industrial significance. However, this was not destined to last for long because already at beginning of the twentieth century the shallowness of the lagoon waters progressively compelled its shipbuilding activities to be relocated elsewhere. A few decades later, in the 1950s, the Arsenale was also deprived of most of its military functions as the Italian Navy decided to transfer its key outpost on the Northern Adriatic Sea from Venice to the city-port of Ancona in central Italy (Lugato and Ruol 1994).

With the swift decline and depopulation engendered by the loss of these vital activities, the Arsenale became a rapidly decaying urban area under the

---

3 The Arsenale covers approximately 32 of the 670 hectares on which Venice is stretched (Chirivi 1983).

administrative ownership of both the Italian government and the navy. This preserved its physical and symbolical detachment from the rest of the city until the 1970s, when the issues afflicting this central area began to be conceived as inseparable from the problems of contemporary Venice. During those years, in fact, Venetian architect Romano Chirivi (1976) with other fellow colleagues of the IUAV set out to work on a new urban plan for the whole district of Castello where the Arsenale is located. As Chirivi recalls (in Grandesso 1988) the key goal of his plan was to draw the attention of the urban elites to the relationship between the Arsenale and the city – a relationship that the Navy had intentionally and constantly attempted to obliterate ever since it took possession of the Arsenale. While his projects very soon ended up forgotten and buried in the drawers of the City Council (Chirivi in Grandesso 1988), Chirivi's purposes were eventually achieved. Not only for the first time in centuries did the political and economic elites of Venice rediscover the Arsenale, but they also begun to develop the perception that an exceptional opportunity was concealed within this monumental area: the opportunity to inscribe the urban space with new functions. This was, indeed, an extraordinary occasion in a city whose spatial organization has always been constrained by the unique presence of water and 'history'.

From that moment onwards, the key question of the re-utilization of a huge urban void lying at the heart of Venice, with its invaluable historical and architectural heritage, engendered a countless number of proposals, initiatives, and debates (see Lombardi and Paternò 1992) especially among experts and local decision-makers. These latter became aware that any future function of the Arsenale would inevitably affect the future role and image of the whole city. Put in another way, the debate prompted by Chirivi was perceived to be of crucial significance for the city. By foreshadowing the prospect, or the threat, of changing its economic settings and social hierarchies, it attracted the attention of the most eminent social actors of the local milieu (Grandesso 1988).

Conceived as such, the Arsenale was not just a monumental building where the glorious past of Venice was passively 'stored'. Rather it was a non-human element that actively triggered the (re)constitution of local networks, mobilized either to propose or to hinder its regeneration. In ANT vocabulary, for its capacity of modifying the state of affairs, the Arsenale was an 'actor' of these networks because it enabled their collaborative alliances, such as those that would be established around the project of re-utilizing this urban space to set up the Center for Marine Technologies.

As recalled in more than one occasion during the ceremonial anniversary of Thetis, this idea was initially conceived in a dissertation carried out at the IUAV, where in the late 1980s the revitalization of the Arsenale was one of the most popular topics investigated by students and academic staff. This interest mirrored the broad and enduring attention paid by the urban elites to the future of this emblematic site. However, notwithstanding the large number of regeneration schemes hitherto proposed, their realization had always been prevented by

Figure 9.1 The main gate of the Arsenale (photo by Mauro Cannone)

the systematic opposition of some elite individuals or groups (Grandesso 1988). Crucial in this regard, the methodological approach employed by three students of the DAEST[4] entailed a fundamental reversion of this common yet problematic procedure: instead of proposing a project and then hoping that it could gather enough support, consensus, and thus resources to be carried out, these students set out to acknowledge at first the opinions of the local elites, with the aim of translating these opinions into ideas and solutions, supported by a visible consensus. As Ms Grandesso (2006, interview with the author), one of the three students involved in the project, explains in more detail:

> In a small city with very limited free spaces such as Venice, the decay of an area covering large part of the built environment was understandably perceived as a real urban problem and the University acted as managerial class in trying to solve this problem ... Our dissertation was in effect one among many other studies that the IUAV proposed to its students at that time. It was divided in two temporal stages: first, we carried out interviews and questionnaires with 44 local actors, representing the business sector, hoteliers, politicians, and intellectuals ... With this wide selection, we sought to understand what the local elite in general had in mind for the Arsenale,

4 The Department of Social and Economic Analysis of the Territory, at the IUAV.

and as a consequence for the city. Three main proposals emerged: the first was related to the creation of a marina for elite tourists, which was the more lucrative solution ... the second was that of an art exhibition space. And then, obviously the third option – which by the way was also the one that gathered lesser consensuses among our interviewees – was that of setting up a center for Marine Technologies.

The innovative methodological framework of the dissertation was suggested by Professor Indovina, the supervisor of the project, who had a very precise idea in mind for the future of the city. To be sure, his idea mirrored a quite widespread opinion in the academic and political environments of Venice, an opinion supported by the awareness that 'one and a half centuries of industrial modernity had compelled Venice to terrible 'out-of-time' conditions' (Fabbri 2005, 28; see also Costa 1990; D'Agostino 2002). On the contrary, under the contemporary paradigm of post-Fordism, the city has concrete chances to redeem herself from these conditions by attracting the new immaterial functions that characterize the development of many other urban centres of the post-industrial world (see Perulli 1989; Rullani and Micelli 1998). Following this general idea, the ambition of Professor Indovina was to trigger the rebirth of Venice precisely as a new pole for these functions. Importantly, this personal belief modified the course of the events, marking a decisive turning point for the future of Thetis:

> I first came to Venice in 1972 [Professor Indovina (2005, interview with the author) recalls] and 3 or 4 years later there was a seminal conference to promote this as the city of 'immaterial activities'. We've been talking about this idea ever since ... and I've never lost my faith that the problems of Venice could be the very chances for her rebirth, her transformation, and that the Arsenale could be the place for this rebirth ... This is the idea: the solutions to the problems of Venice are very complex, but they can also be sold to the rest of the world ... These are the real treasures of Venice! For example, the problem of safeguarding the city from floods, in the next 20-30 years will certainly become a problem that other cities, and not only Venice, will have to face ... And the Arsenale can be the center of all these activities, it should become the centre of scientific research and excellence, something linked to the private sector and at the same time to the University.

On the completion of the first phase of their research, the students concluded that the opinions of the urban elites could be generally drawn together in three key ideas (as it has been mentioned above). The second phase of their work consisted of further investigation into each of these ideas. Antonietta Grandesso, the only Venetian among the three, was selected by Professor Indovina to focus on the project of forming the Center for Marine Technologies, which was the kind of solution that Professor Indovina had in mind for the revitalization of the Arsenale and of the city. In setting out her individual analysis on the future repercussions that such a centre would have triggered in Venice, Ms Grandesso also explored the ways in which this initiative could be practically implemented, and in doing so she worked closely with all those institutions that had previously proposed the idea of the centre. In concluding her work, she argued that 'the abandoned warehouses of the Arsenale were the spaces in which the renowned marine technologies of Venice had been developed for centuries. So why don't we use these spaces to produce these technologies as in the past?' (Grandesso 2006, interview with the author).

Importantly, the choice of this particular option produced an initial definition of a potential set of actors that, thanks to the work carried out in the first phase, were supposedly interested in the project. This is what in ANT vocabulary is known as 'problematization', namely 'a form of translation that posits an equivalence between two problems that requires those who wish to solve one to accept a proposed solution for the other' (Callon et al. 1986, XVII). In fact, the solution proposed by Ms Grandesso and her supervisor aimed to regenerate the Arsenale and trigger the socio-economic revitalization of Venice. The idea of the Center for Marine Technologies was also particularly attractive for some public and private institutions, as its creation would have allowed them to further develop their institutional objectives. As a consequence of this common interest, the question, 'Why don't we use the decaying warehouses of the Arsenale to produce marine technologies?' (Grandesso 1988), set the ground for the establishment of new alliances among some social and material actors of the local milieu. More precisely, it did so by providing an initial 'interdefinition' (Callon 1986) of these actors.

The 'Interdefinition of the Actors'

Following the results achieved in the first phase, the aim of the second part of the dissertation was that of analysing the consequences that a Center for Marine Technology located at the Arsenale would have implied for Venice. But in her attempt to anticipate these consequences by means of interviews, questionnaires, archival research and similar methodological tools, Ms Grandesso provided a first identification and classification of three socio-material entities. And in carrying out her research project she established unprecedented associations and links between them. More precisely, these actors were:

a) *the Arsenale*: described in the dissertation as a monumental area of immense historical and cultural importance, which at the time was decaying as a result of the significant costs of restructuring and maintenance, and above all because of the internal disagreement among the urban elites on its future destination.
b) *the urban milieu of Venice*: comprised of the 44 interviewees of the dissertation, chose as the spokespersons of all the different categories, groups, and organizations acting in the city. The different solutions that these interviewees first proposed to the problematization of the Arsenale were successively grouped together in the three main ideas that Ms Grandesso outlined above. By associating each spokesperson with

one of these ideas, the dissertation divided the urban milieu of Venice into three groups, creating a virtual opposition among them, but also producing a potential cohesion among the actors of the same group.

c) *the group of actors proposing the solution of the Center*. Among the three ideas proposed, some individuals and organizations shared the same specific interest of creating the Center. In particular, this group of actors was composed of private and public institutions which together constituted a peculiar socio-economic character of the urban milieu, namely its long-standing expertise developed in the field of marine technologies. The spokespersons of the organizations where this local expertise was produced – namely the spokespersons of research institutions such as the CNR and of private companies such as Fincantieri and Tecnomare[5] – perceived the idea of the Center as that of a unique opportunity to merge their different knowledge and missions and thus as an occasion to pursue their institutional interests.[6]

The solution to the initial 'problematization' of the Arsenale established a fundamental correspondence between the research project, the interests of the actors above described, and that of preserving and regenerating the Arsenale (see Figure 9.2). Moreover following Professor Indovina's idea, the solution of establishing the Center for Marine Technologies acted as an empirical occasion and stimulus to transform the exclusive characters of the urban milieu into resources of local development. For Emanuel and Governa (1999), the contemporary dynamics of urban change are enacted precisely when, as in this case, the unique characters of the urban milieu are valorized as potential resources of territorial evolution, and when there are projects setting in motion the effective translation of these preconditions.

In this respect, the project outlined by the research team stimulated the establishment of a local network of socio-material actors associating to work

5 The CNR (Consorzio Nazionale delle Ricerche, the National Research Council) is a public organization whose duty is primarily that of promoting and carrying out research activities in various domains of knowledge (mainly: scientific, technological, economic and social). In Venice, the Council was structurally organized into two different institutes focusing on marine science studies and on environmental processes dynamics. Tecnomare and Fincantieri are two of the very few private industrial enterprises still located in Venice: the former operates in offshore design and engineering services and it is owned by ENI, the largest Italian public company in oil and gas sector; the latter is a state-owned company operating in shipbuilding industry (Distefano and Paladini, 2002; see also http://www.cnr.it; www.fincantieri.com; www.tecnomare.it).

6 As Grandesso (2006, interview with the author) remarked, direct forms of cooperation between public research institutions and private companies were not allowed under Italian law, and as a result, the proposal of constituting a centre in which such cooperation could be indirectly but legally carried out was perceived as an ideal solution.

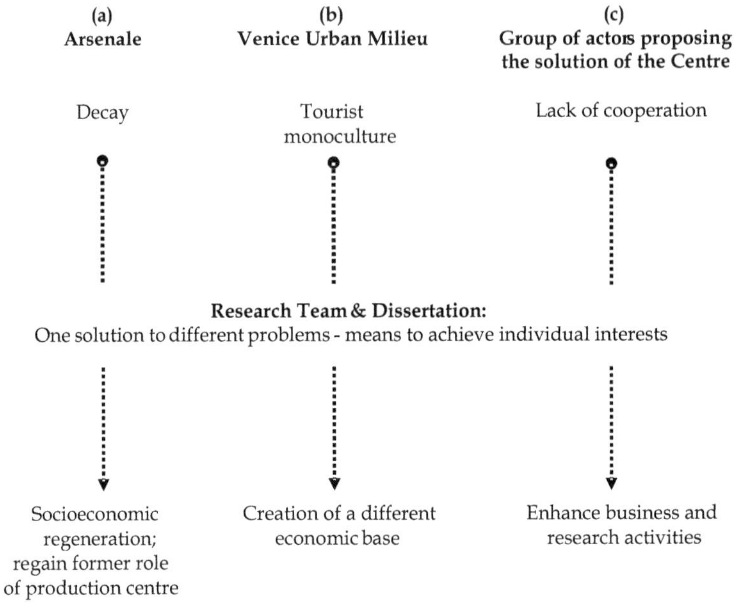

Note: From top to bottom: the actors identified in the dissertation; their problems; the central role of the research team and of the dissertation in suggesting a common solution to all these problems; the specific interests that each actor could achieve only by submitting to the set of alliances proposed by the research team in the dissertation

Figure 9.2 The set of alliances proposed by the research team

Source: elaborated from Callon 1986.

on the implementation of a key idea. More precisely, it did so by establishing a clear symmetry among them, or, more precisely, among their interests. Not only were all these heterogeneous actors mobilized by the questions posed in the dissertation – 'the Arsenale: towards what destination?' and 'what are the consequences of the constitution of a Center for Marine Technologies at the Arsenale?' (Grandesso 1988) – but crucially, they were induced to acknowledge that their cooperation around a shared task could benefit them all.

As shown in Figure 9.3, the common solution to different interests required, in the first place, the approval of the virtual set of alliances proposed by the dissertation, because the alternative for all of these actors was that of continuing to act independently, with all the difficulties that this entailed. But in accepting to work together, the research team, the Arsenale, and part of the urban milieu would have found themselves locked into a set of associations necessary to overcome, at once, all the obstacles to their particular interests. In this way, as Ms Grandesso's words suggest, while different identities and goals begun to intertwine with one another, 'social' relationships between different actors started to acquire the guise of a network with a common objective to achieve:

> In doing this research, I started to develop not only a deep interest in the problems of the city, but also an enthusiastic attitude to work on their solution. I realized that I was accidentally building an informal network, a network of communication between local private companies and public research institutes, and real consequence of my work was the practical proposal of setting up this Center ... All the people I've worked with were very positive because there was, and I discovered it, a growing interest to create this Center, to practically get things done. (Grandesso 2006, interview with the author)

By acting as a 'mediator' (Latour 2005) between – or as 'spokesman' (Callon 1986) of – social and material entities which were previously almost utterly unconnected, she progressively convinced them to work together, while their languages, aims, and resources were 'translated' in her academic work, into a first step towards the constitution of the Center. While carrying out her research project, she interacted simultaneously with the different institutions selected, exercising on them a 'power of persuasion' that for John Allen (2003, 123) is among those associational powers crucial 'to connect, to bring together but not to suppress the interests and differences that commonly divide'. In addition, it seems that this capacity was also the consequence of hierarchical settings that reversed the common rules of *power*, at least in its mundane understanding. The very fact of being a student and thus of not being 'powerful' in their perception, endowed Ms Grandesso with the strongest persuasive power she could have ever exercised over them:

> In my view, almost half of the reason of this success was due to the fact that I was a student. If I'd do the same now [they would ask] 'What does she want?' 'What does she want to steal?' They would see me as a competitor, as an enemy ... On the contrary, I was perceived as a very positive link among them, because I was not a competitor, but just someone who was talking to them ... A young student does not frighten anybody, and so people were actually very friendly, because they saw the enthusiasm of a young person, they saw 'innocence', and in these cases even the bureaucrat is unlikely to give you hard times. (Grandesso 2006, interview with the author)

What would have been the result if anybody else in a different 'social' position had tried to do the same work of 'mediator' between these public and private institutions is a question with no certain answers. What seems to be more certain from her account, however, is that mutual trust and cooperation did not appear spontaneously as by-products of 'history', in spite of Putnam's (1993) idyllic observations on the Italian region where Venice is located. Rather, these social resources emerged gradually as a consequence of her connective work of *negotiation* (Latour 2005, 39) between the particular aims of the institutions in which she carried out her research, and the general interest of setting up the Center. In such cases, the alliance among different actors around a proposed problematization is strengthened and modified all along a process in which a virtually unlimited variety of social mechanisms, ranging from pure force to seduction or solicitation, can be deployed (Callon 1986, 209). During her research, presumably with a certain behaviour and attitude, she established a constructive predisposition towards herself as 'mediator' or 'spokesman' of others: 'accidentally', as she claimed, she shaped their will so as to make them follow her project. She built an informal 'network of communication' by projecting the existing cooperation for her research assignment as a desirable way to work for the future constitution of the Center. In this way, she strengthened, or built anew, a 'capital' made of mutual acquaintance(s), relationship(s), and recognition among a specific group of actors operating within the local milieu, a capital which could be spontaneously enacted as a fundamental resource to translate her idea in reality:

> All the actors I have managed to put together cooperated in the correct way, namely with no competition but rather with a 'united we stand, divided we fall' sort of attitude. But I also believe that, especially at the very beginning, this network was absolutely precarious … I mean: they said that yes, they were interested, but probably because it was not such a big commitment. It was, in fact, just research, a study … But during this research we developed a very effective method of working together, and this probably convinced them to go ahead with our project because our cooperation was working. Also, we tried to promote some sort of empirical action which was not necessarily associated with any political 'faction', and in this way we could mobilize our social relationships in the city without any major opposition. (Grandesso 2006, interview with the author)

In promoting empirical 'action', the academic partnership between Ms Grandesso and her supervisor progressively established itself as an 'obligatory passage point' (Callon 1986, 205) in the network of relationships that was being built. In other words, they established the perception that the set of alliances they were proposing within the city had inevitably to pass through and be managed by their academic partnership, which for this reason became the focal point of control of the network constitution. And in this way, the dissertation acquired the role of a device of *interessement* luring any potential actor to submit to this set of alliances, and strengthening the solidity of the network further.

The Device of 'Interessement'

According to Callon (1986, 207–8) *interessement* is a crucial mechanism in the act of network building. 'To be interested', he argues, 'is to be in between (inter-esse), to be interposed ... To interest other actors is to build devices which can be placed between them and all other entities who want to define their identities otherwise.' In order to strengthen a hypothetical alliance around a given objective, an actor should weaken the links that its potential ally could establish with a third party to achieve the same goal.

In this case, the 'interessement device' took a very material form. When he was interviewed as one of the 44 privileged informants selected in the dissertation, Domenico Lalli was the CEO of Venice's Tecnomare. Remarkably, he was among the main supporters of the idea of the Center for Marine Technologies:

> My opinion of the Arsenale was that of a geographical area of the city particularly ideal to develop those marine technologies focused on sea pollution, flooding, tides, and all those problems that in Venice we know very well. That's why at the very beginning, when I was interviewed, I suggested this idea. Then I saw the dissertation, and I saw a chapter entitled, more or less: 'The revitalization of the Arsenale as the area of marine technologies' where ...where it was proved not only that this idea was in fact shared also by others, and therefore that there was a sort of consensus around it, but also that it would have had positive socio-economic effects on the city ... We had this coincidence of ideas and interests, and ... that's it! ... That's why after Antonietta [Grandesso] had completed also the second part of her research I was even more convinced that the idea proposed in the dissertation could be effectively implemented. This idea was like a match that started a fire. (Lalli 2006, interview with the author).

In this account, the dissertation appears as a material device of interessement 'responsible' for spreading such fire, because it proved to Mr Lalli that his idea was also shared by other actors of the urban milieu, suggesting in this way, a hypothetical alliance with them and with the researchers, while weakening any potential intention to achieve the same goal independently or with third parties. 'The interessement [in fact] helps corner the entities to be enrolled ... it attempts to interrupt all potential competing associations and, as a result, social structures comprising both social and natural entities are shaped and consolidated' (Callon 1986, 211).

These effects were produced by the very materiality of the device, able to form a relatively stable network by fixing its virtual associations in a durable material, thus providing them with a 'steeliness' that they would not have possessed otherwise (see Law 2001, 6). In this respect, the dissertation established itself as an indispensable tool to bring the interest of Tecnomare closer to the solution of the research team, and it did so by exercising a power of persuasion that eventually convinced the company and its CEO to accept such solution. This persuasive power was evidently exercised when the conflictual or cooperative links among 44 local actors were transformed from ethereal thoughts and opinions into visible words, by virtue of which these individuals and groups became either connected with, or separated from, one another in a durable fashion. As Law (2001, 6) argues, in fact, because of their ethereal character, 'thoughts are cheap but don't last long, and speech lasts very little longer. But when we start to perform relations – and in particular when we embody them in inanimate materials such as texts and buildings – they may last longer'.

In this case, the act of gathering information, recording, and charting words and opinions, extended and imposed an unprecedented transparency and order in the social milieu of Venice, because it shed light on what could be done, and by whom, to solve the problematization of the Arsenale and thus to contrast the socio-economic decline of the city. Both the consensus and the disagreement between various actors could therefore be used to administer this problem in an objective way, precisely because thoughts and opinions took a paper form. And when 'statistics, numbers, classifications, judgments, drawings, designs, records and files' (Allen 2003, 132) take a solid form, the people aligned behind them do not melt into air, and their relationships do not need to be continuously renegotiated and re-established (see Callon and Latour 1981, 296; Latour 2005, 68–70).

In short, the urban milieu of Venice was broken down into 44 spokespersons to be interviewed and then regrouped on the basis of their opinions. The groups, opinions and ideas were made easy to transport and visualize, but difficult to contest and renegotiate. The dissertation ascertained and showed as a 'matter of fact' that the realization of the Center for Marine Technology was virtually possible because there was an 'objective' expertise within the local milieu to realize the Center, and also that there was a 'subjective' consensus within its social environment towards such a project. Recalling Lalli's metaphor, Ms Grandesso's work on paper was likewise a match that started a fire across specific buildings, offices and institutions of Venice by showing the empirical evidence that the preconditions to set up the Center were real, because they were supported by statements and opinions recorded 'objectively' and because all the preliminary discussions and negotiations among local actors had already been carried out.

By virtue of the particular business prospects that the dissertation prefigured for the company, Tecnomare was the first place hit by this fire: in Callon's (1986) vocabulary, Tecnomare became an actor whose 'interest' was elicited by the 'problematization' advanced by Ms Grandesso. Importantly, this interest produced a decisive effect when the company's CEO decided to commission a feasibility study for her project, under the direction of the same supervisor. This request marked another turning point in the constitution of Thetis. As with the factual enrolment of Tecnomare, the associations between different elements of the network also began to be performed *outside* their paper form and *outside* the academic environment of the IUAV: they begun to be projected and assembled in the wider social and spatial dimension of the urban milieu, where the project could get hold of the empirical resources essential for its final implementation. In other words, while in the first phase the successful partnership that had originated at the IUAV blended the knowledge and particular ideas of Professor Indovina with the enthusiasm and research skills of Ms Grandesso into a powerful social symbiosis, now these intellectual resources were about to be joined with other fundamental qualities of *capital* value possessed by Tecnomare and by its existing 'allies'.

The 'Enrolment' and 'Mobilization of Allies'

> I was very confident that I could give an important contribution to the project, not only because I was working in the same sector of marine technologies, but also and more importantly because I had practical experience as to how good ideas could be translated into reality, even if I had to start completely from scratch. After all, I did the same when Tecnomare was established … and therefore I was confident to be able to manage this potential idea of Thetis and to be able to mobilize the resources and the people that could potentially play a decisive role in translating it into reality. (Lalli 2006, interview with the author)

'To mobilize', as Callon (1986) observes 'is to render entities mobile which were not so beforehand'. The enrolment of Tecnomare entailed its effective mobilization, that is, its *transformation* from passive element of a virtual network into an active participant in the process of network building. In this new guise, Tecnomare took the role of the principal 'ally' of the collaborative association between Ms Grandesso and Professor Indovina, because the company made available not only its own financial resources for the feasibility study, but also a much wider set of resources ranging from the crucial expertise of its CEO to the various forms of support that could be provided by all its existing business partners.

In this particular respect, social relationships were a fundamental tool to further mobilize individuals and organizations and access their specific resources to set up a Center for Marine Technologies. Mr Antonio Paruzzolo, for instance, was a former manager of Tecnomare living and working in the US at the time during which his previous company was embarking on the adventure of Thetis. Aware of his managerial skills, but also of his wide credentials in terms of reliability and trustworthiness, his former Italian employers contacted him for the job of managing the new Center that was being constituted. And the same basic idea of deploying existing acquaintances, friendships, and relations to reach out for potential allies with different resources – from finance to knowledge, from entrepreneurial to organizational skills – was applied in the successive phases of constitution, when these *social* relations became an indispensable *capital* to translate an academic study in reality.

As Mr Paruzzolo argues, there is little doubt that the particular interests of all the potential allies played a major role in this process of resource mobilization. However, Mr Paruzzolo – who was one of the key protagonists of this process – put such interests in a much wider 'social' frame, recalling equally significant details, coincidences, and facts that played in favour of the successful results eventually achieved.

> Our relations and acquaintances played absolutely a key role. Without these specific relations in that given time, we wouldn't be here today. More than money, more than knowledge, more than everything else, it was a matter of who knows whom … For example, Professor Indovina was acquainted with Professor Puppi … who at that time was the president of Tecnomare … In this way Mr Lalli was informed of the dissertation … and then he contacted Tecnomare's business partners for the feasibility study, assuming that they could be interested in expanding their industrial activities in the marine technologies sector. … We identified 5 or 6 industrial partners, and together we set up a consortium with the aim of working together at the Arsenale. (Paruzzolo 2006, interview with the author)

The partnership that translated Thetis from a theoretically grounded academic idea into a concrete business and research consortium was thus also triggered by the enactment of pre-existing relations, such as the personal acquaintance between two crucial local actors representing the academic (Professor Indovina) and the private sectors (Professor Puppi) of the city. Subsequently, the process of enrolment of different local enterprises, as Mr Paruzzolo suggests, was set in motion by virtue of existing working relationships, and translated according to the ways in which the specific business domains and interests of these enterprises could be negotiated in relation to the original aim of the dissertation. Certainly, this was a process that developed interactively, by bringing together different networks and anticipating how the potential collaboration between them could work, by discussing what, in effect, this new set of partnerships could produce.

The institutionalization of this research and business organization was eventually signed in 1993, when Thetis was formally registered as a consortium funded and controlled by industrial companies, research institutes, and political institutions, with the aim of developing technologies and supplying projects focused on the marine and coastal sector (Distefano and Paladini 1997). However, the key goal with which the idea of Thetis was first conceived – the regeneration of the Arsenale – had yet to be achieved. In those years, the Arsenale was still a semi-abandoned military area, largely inaccessible because of the presence of the Italian Navy. As a consequence, the consortium had to settle elsewhere in central Venice while it worked to obtain the concession required to relocate to the Arsenale. Importantly, this very task required the transformation of the initial composition of the consortium, as Mr Paruzzolo explains:

> Obviously, as long as we were dealing just with marine technologies we didn't need political relationships. But these relationships became necessary when we decided to regenerate the Arsenale, because we had to obtain the concession and the funding, 6–7–8 billion lira, to conclude our project ... And that's how a new phase of Thetis got started: in that moment the goal was ... that of finding the sponsors of this project, which meant finding the partners financing its total cost. Thus, the initial structure of our consortium could not hold any longer ... and our motto switched from 'let's bring ideas together' to 'let's bring money together'. We kept some of our old partners, but we also had to find new ones by 'asking money to those who have money', as the proverb goes. And to do so we mobilized our existing contacts, which means that we don't deal anymore with, say, the technicians but with the CEOs of some of the major business groups in Venice and in Italy (Paruzzolo 2006, interview with the author).

The socio-material network of Thetis was ready at this stage to stretch further towards new associations. While its action had hitherto been enclosed within the Venetian milieu where it progressively enrolled actors, groups, and locally embedded resources, it eventually required new connections to the supra-local level to achieve its most symbolic goal.

And it was precisely the powerful symbolism concealed behind the intention of rejuvenating the Arsenale that enabled the mobilization of new 'allies' with resources that necessarily bypassed the local political and economic environment. The general principle promoted by this initiative, was in fact, that of 'innovation' – of the city, its spaces, functions, and social composition – which for various reasons large parts of the local elites had hitherto successfully opposed with a pervasive attitude of conservation (Bonomi 1995; Sbetti 2005). As Dorigo (1973) has discussed, over the last two centuries the preservation of the built environment has been deployed as a powerful political and economic tool for the exploitation of the 'stones of Venice', as Ruskin would say (for a detailed historical account, see Dorigo 1973; Ginsborg 1978; Dri 1993).

Following this general trend, the urban core has been studiously protected, even during modernization, which in Venice coincided with the development of industrial and leisure functions directed specifically toward the edges of the lagoon, with the creation of the industrial and residential clusters of Marghera and Mestre on the mainland, and the constitution of the Lido island as sea-resort on the opposite end (see Fontana 2002). Gradually abandoned by the activities and inhabitants attracted to these more peripheral areas, the spaces of central Venice have been freed and then 'frozen' as tourist simulacra. Against Marx's famous statement concerning the effects of capitalism on urban spaces, in the last two centuries – and especially in the last 50 years – in Venice all that was solid *did not* melt into air, because instead of 'transformation' here the key word for capitalist profit was 'safeguarding' – a word too often employed to disguise private interests as disinterested love for the city. A word too often employed just as a rhetoric tool to contrast *a priori* any idea and project of innovation, and to promote the unconditional preservation of the urban environment for its complete devotion to tourist exploitation and land revenues (D'Agostino 2002; Dorigo 1973; Paladini 2002).[7]

Arguably, the successful implementation of the Thetis initiative could be perceived as a threat by many supporters of this conservative ideology, because it incorporated the risk of uncovering the inconvenient truth: that the present and the future of Venice were by no means predetermined by a romantic fate, but rather, driven by a precise will not to change the current state of affairs (on the long-debated conflict between conservation and innovation in Venice, see Distefano and Paladini 1997; Dri 1993; Dorigo 1973; Isnenghi 1997; Mazzariol 1985; Paladini 2002). As a consequence, the Thetis initiative symbolized a clear challenge to particularistic interests, a very difficult challenge that required the substantial support of 'multi-level' local actors (Emanuel and Governa 1999) and the mobilization of their resources, as the CEO of Thetis recalls:

> In trying to obtain the concession of the Arsenale, Thetis acquired a new meaning, a socio-political meaning. In those years, Thetis became the feather in the cap of the urban administration, and so the mayor became the intermediary of a long list of relations that in some ways supported our project. That is why we were able to obtain funding from the EU, from the Regional and Urban governments, and from various business partners, both old and new. In those 5 years our target was that of completing the renovation of our concessionary buildings at the Arsenale, and we got that done thanks to these funding opportunities and contacts ... For example, a person without whom this project wouldn't have been possible was the CEO of ENI, who was our missing link: we were short of 3.5 billion lira, and he decided to put this money in our project. ... I got in touch with him through the mayor, who knew him better than us. (Paruzzolo 2006, interview with the author)

A few days before the official opening of the laboratories and office spaces of Thetis at the Arsenale, in his keynote address at the conference 'Working in the net', Mayor Massimo Cacciari launched a clear appeal to the social and economic groups of his city, all of which were represented among the audience: 'Either Venice will be able to attract enterprises ... working in the new technologies sector, or this city will become a ghost dominated by the tourist monoculture' (quoted in Distefano and Paladini 1997). After the relentless decline of her industrial activities, as far as Cacciari was concerned, Venice required urgent economic revitalization to challenge the deathly spectre of

7 A principle that the national government institutionalized in the 'Special Laws for the Safeguard of Venice', according to which anything *new* had to be built in Mestre and Marghera while 'not a single brick of the [historic] city could be removed' (Dorigo 1973, 82). Since the 1950s, this law has been utilized as a strategic tool by quite a diverse social and economic alliance, from the industrial and residential developers of the mainland to the business group and individual enterprises operating in the tourist industry and in the real estate sector (De Michelis 2005; Dri 1993, 44; Indovina 1995; Paladini 2002).

tourism. As mentioned above, rather than just a personal belief, this idea was part of a much wider political and cultural project put forward in the 1980s by a group of intellectuals, academics and political leaders, acting in both the context of the two universities and local political and cultural associations such as the civic list Il Ponte and the Istituto Gramsci Veneto (Fabbri 2005; Paladini 2002; Sbetti 2005). This project materialized in 1990 in the guise of an edited collection proposing a comprehensive new Idea di Venezia (Cacciari 1990), with the aim of suggesting innovative 'ontological solutions' for a city that, during the contrasted transition from romantic decadence to industrial modernity, had lost her peculiar identity as well as the capacity of self-representation (Fabbri 2005; Sbetti 2005). Eventually, this project and its publication became the theoretical and political manifesto supporting the electoral campaign of the undisputed leader of the group, Venetian philosopher Massimo Cacciari, who was officially elected mayor of the city for the first time in 1993 (Fabbri 2005, 11).

As Bonomi (1995) argues, this election determined an 'ideological' change of direction for Venice, because the political project of the new urban government was that of re-addressing the utter conservativism of the last two centuries by promoting an idea of Venice as a city which had to preserve its extraordinary uniqueness while developing the same economic functions of other more ordinary urban centres (see Cacciari 1990). Within the new political agenda, the Arsenale was the ideal location to stimulate such a long-awaited process of urban renewal (see Costa 1990), and for this reason the Thetis project became the 'feather in the cap' of the urban government, as it symbolized the same paradigmatic criteria of innovation advanced in the political manifesto of the mayor.

However, notwithstanding this virtually ideal combination of political and economic objectives, the association between these two very influential networks of local actors came as the result of a twofold movement of negotiation, in which their parallel projects eventually intertwined becoming mutually indispensable. The enrolment of further 'allies' and resources that *only* the urban government could mobilize was essential for the Thetis network just as much as the regeneration of the Arsenale, and the way that *only* Thetis could showcase it, was essential for the urban government to trigger the renewal that it sought to promote.

This twofold dialectic movement between the economic and political components of the milieu implied that, on the one hand, the city council had to mobilize its own networks of contacts to access the capital resources that Thetis required, while on the other, the project leaders of Thetis had to literally 'take on board' the new symbolism attached to their project, modifying their mission and organizational structure so as to preserve the interests of the new actors to be enrolled. In turn, this implied another crucial *translation* in which the regeneration of the Arsenale had to be performed so as to take into account this set of relationships with new political and economic allies. Only in this way, in fact, could the mission of reconverting the Arsenale back to its former productive function become the declared objective of all the actors associated in these networked relationships, because, as Isin (2002, 42) observes, echoing Lefebvre (1991) 'the creation and production of space' is a crucial tool through which social networks 'bring themselves into real existence'.

However, while this set of associations became visible through the recreation and reproduction of the Arsenale, it was the materiality of the Arsenale that produced their (networked) relations and that held them together in space and time. The Arsenale in all its decaying materiality, allowed their interests to converge towards its regeneration and their resources to be accessed and transformed into capital to achieve this shared objective. As Latour would suggest (see quotation at the beginning of this section), if for a moment we erase this material character from the Venetian landscape, the objectives, interests, resources, and networked relations among these actors would disappear from view together with the Arsenale. After all, aside from a number of other reasons, the simple fact of its existence, of its being 'there' as a symbolic space and a monumental and historic site of political control, caused a series of 'social' effects in space and time. It 'provoked' the interest of the IUAV, enacting the intellectual resources of Professor Indovina and Ms Grandesso to form an effective 'social' cooperation and in turn this cooperation produced a project that acquired capital value for Tecnomare, which eventually mobilized its knowledge, ideas, expertise, contacts across the local milieu to translate this project into reality. To develop the Arsenale, an endless number of interactions and interests that are typically defined as 'social' were therefore necessary, while these social elements also had to intertwine and be preserved in a material project. This project could thereby only become real insofar as all the interests of its social protagonists were preserved by intersecting the one with another, because 'if the range of interests gathered around a project do not intersect', and if 'the two-way movement translating interests and modifying the project is interrupted, then a project cannot become real' (Latour 1993, 391).

Conclusion: Social Capital and the Socio-material Dialectic of Urban Regeneration

Today, after many more exchanges, translations, and negotiations on top of the ones that I have only partially described here, Thetis is a private shareholder company working primarily on the environmental problems of Venice and successfully using this 'local' expertise as a unique capital to exchange on 'global' markets. Today, Thetis employs about 180 people who, against the current process of depopulation, go to Venice to work and sometimes even to live. In this way Thetis contributes to the regeneration of the Arsenale (and of the city) not only by showing symbolically that this can be a pulsating heart of the contemporary knowledge economy, but by also being one of the key institutional actors currently working on such an ambitious project.

Figure 9.3 External view of abandoned warehouse in the Arsenale (photo by Mauro Cannone)

Today however, behind the materiality of Thetis' 10,000 m^2 outdoor area, behind the 5,000 m^2 of old warehouses converted into office and laboratory space, behind the technological instruments worth €6–7 million (Paruzzolo et al. 2007), some fundamental yet invisible resources and processes are concealed: resources such as the intellectual capital of Professor Indovina, the enthusiastic attitude of Ms Grandesso, the interest of Tecnomare and of the local business community, the entrepreneurial skills of Mr Paruzzolo and the Idea di Venezia of local political and intellectual leaders. And then there are the 'social' processes, such as the ones with which these local resources have been invested and transformed to create an occasion of 'development'. Precisely because they are unseen, these resources and processes need to be unveiled and recognized. Otherwise one could argue (in a rather straightforward fashion) that Thetis and the socio-economic growth that it contributes to are the result of a 'social capital' of civic traditions and cooperation that makes the mechanisms of the economy work. Perhaps even more importantly, they also need to be unveiled and recognized to show that although the future of Venice *might be* that of becoming a theme park, of being 'saved' by a corporation, or even of being allowed to die, none of these options is yet determined as an inescapable matter of fact. On the contrary, Thetis demonstrates that this city has concrete opportunities to set herself free from such problematic issues as her dependence

Figure 9.4 Internal view of an abandoned warehouse of the Arsenale (photo by Mauro Cannone)

on the tourist sector and dramatic depopulation, which jeopardizes her urban functions.

In other words, not recognizing the significance of the presence 'in place' of specific resources, and of the ability to mobilize, manage and make an effective use of them, could lead one to believe that socio-economic development has no connection whatsoever with the territorial preconditions *differently* embedded in every local milieu, or the ways in which these potentialities are enacted by networked relations (see Allen 2003). In this case it could lead one to 'forget' the value of the Arsenale (one of the most symbolic attributes of the Venetian landscape), the local university (with both its active managerial role in the city and its more customary function of producing knowledge, ideas, and projects), and of Tecnomare (heralding, with other public and private institutions, the specific expertise developed in Venice within the field of marine technologies). Moreover, it could lead one to 'forget' not only that these potentialities and their presence *in place* were essential to create Thetis, but also that they had to be activated, valorized, and translated into effective resources. And this is where the factual significance of the 'social' as a 'capital' played a decisive role: it was exactly by means of a series of *social* relations that these potential preconditions were transformed into key resources, or a *capital*, for development.

As I have tried to show, each of these resources acquired capital value only when they were put in mutual relation, that is, only when they were associated with one another and made accessible within a socio-material network of local actors. By themselves, the ideas of the research team of the IUAV, the local expertise in marine technologies, the organizational experience and business contacts of Tecnomare, the reputation and acquaintances of Mr Paruzzolo, and the supra-local contacts of the urban government would have been of little or no use or, in Latour's (1986) and Allen's (2003) words, they would have remained 'power *in potentia*'. But all these qualities, abilities, information, expertise, contacts, funding opportunities and so forth were identified and then accessed and utilized by means of social relations, and in this way they became a capital for the creation of the Center, i.e. they became 'power *in actu*' (ibid.).

These tools and resources could be utilized as such only because of their place-specific value, namely because they were firmly rooted and produced within the local milieu. As a resource flowing through social relationships (Bourdieu 1986; Coleman 1988; Putnam 1993), social capital was, and is, therefore embedded and enacted within the same 'space' in which networked social interactions were, and are, (re)produced. Namely, on a geographical dimension where individuals associate because of their physical and socio-cultural proximities, a shared sense of belonging, and common identities. And if it contributes to local 'development', it does so not by passively predetermining the ways in which invisible 'societies' act by virtue of their similarly invisible 'histories' or 'cultures', but by facilitating (or hindering) the concrete ways in which real resources could be mobilized to achieve a project that fosters actual 'development'.

As Foley, Edwards and Diani (2001, 277–8) observed, social capital acquires an active role only when 'individual or collective actors ... perceive that some specific resource is present within their social field and ... have some form of social relationship that provides access to these resources'. The constitution of

Figure 9.5 The walls delimiting the southern perimeter of the Arsenale (photo by Mauro Cannone)

Thetis was the result of an extensive process of translation enacted precisely by the perception that specific resources were available and could be accessed by means of mutual acquaintance, friendship, recognition and other similar tools of legitimization, utilized by individual and collective actors, to pursue a common objective. The academic intuition of Ms Grandesso, for instance, would have been of no use without her acquaintance with Professor Indovina (and vice-versa), while his contacts with Tecnomare allowed their idea to become an empirical project. Tecnomare's business partnerships were, in turn, fundamental to enroll the economic capital required for the institutionalization of the first consortium, which by attracting the interests and the attention of the urban government, established a decisive alliance for the final step of the project: the return to the Arsenale. In this way, individual or group resources were mobilized by means of social relationships, becoming 'collective' properties of the network structure in which such relationships were progressively being shaped. And for their crucial role of enacting these resources, social relationships became themselves a capital without which the network structure could not have achieved its objective.

For this reason, 'network' and 'milieu' are the two fundamental metaphors with which the concept of social capital can be conceptually re-cast *in space and in society*, and with which its factual contribution to conditions of 'development' or 'underdevelopment' can be re-addressed while avoiding any discouraging determinism. Indeed, the success of Thetis, cannot be fully understood without employing these metaphors and without acknowledging the elements to which they draw attention: namely, to crucial elements such as the potential resources available *only* in Venice – the Arsenale, the IUAV, the public and private institutions and all the other peculiar qualities of this milieu – and to the way in which these qualities were enacted and mobilized through the 'power of associations' (Latour 1986) – the power to connect, to bring together resources and individuals in heterogeneous networks.

The story of Thetis, in other words, is also a story of social capital, although not of an idyllic cooperation stemming from a cultural attitude spontaneously reproduced in history and statically embedded within geographical administrative borders. Rather, it was the result of a process of territorialization (Turco 1988), in which a local 'society' has been successively (re)created *together with* the materiality of its environment.[8] It is a story of territorial development which crucially depended on what potential resources were available and how they were effectively mobilized and transformed by the associational power of social relations. It is, I would argue, a story of an ongoing process of enrolment and enactment of 'the actual or potential resources ... linked to possession of a durable network of ... relationships' (Bourdieu 1986, 248) – a network whose durability and effectiveness, as it has been shown, was solidly cast in, and performed by, relational associations of power which were, at once, both 'social' and material.

Bibliography

Allen, J. (2003), *Lost Geographies of Power* (Oxford: Blackwell).

Aymard, M. (1991), 'Strategie di cantiere', in A. Tenenti and U. Tucci (eds).

Bagnasco, A. (1999), *Tracce di comunità* (Bologna: Il Mulino).

Barbiani, E. (2002), *Cantiere Venezia* (Venice: Marsilio).

Bec, C. (2003), *Venezia. La storia, il mito* (Rome: Carocci).

Bellemo, F. (2007), 'Thetis, calamita di una Venezia finalmente hi-tech', *Gente Veneta*, 10 March 2007.

Benevolo, L. (1996a), 'Quale idea per Venezia', in L. Benevolo (ed.).

Benevolo, L. (ed.) (1996b), *Venezia il nuovo piano urbanistico* (Bari: Laterza).

Bianchin, R. (2006), 'Venezia nel 2030: una città vuota. Niente abitanti ma solo turisti', *La Repubblica* 2 August.

8 Following Law and Mol (1995, 274) when we 'look at the social, we are also looking at the production of materiality. And when we look at materials, we are witnessing the production of the social'.

Bijker W.E. and Law, J. (eds) (1992), *Shaping Technology/Building Society: Studies in Sociotechnical Change* (Cambridge, MA: MIT Press).
Bonavero, P., Dematteis, G. and Sforzi, F. (eds) (1999), *The Italian Urban System. Towards European Integration* (Aldershot: Ashgate).
Bonomi, A. (ed.) (1995), *Privatizzare Venezia: il progettista imprenditore* (Venice: Marsilio).
Bourdieu, P. (1984), *Distinction: A Social Critique of the Judgement of Taste* (Cambridge, MA: Harvard University Press).
Bourdieu, P. (1986), 'The forms of capital', in J. Richardson (ed.).
Cacciari, M. (1990), 'Idea di Venezia', in Fondazione Istituto Gramsci Veneto (ed.).
Callon, M. (1986), 'Some elements of a sociology of translation: Domestication of the scallops and the fishermen of St Brieuc Bay', in J. Law (ed.).
Callon, M. and Latour, B. (1981), 'Unscrewing the Big Leviathan – how actors macrostructure reality and how sociologists help them do so', in K. Knorr Cetina and A. Cicourel (eds).
Callon, M., Law, J. and Rip, A. (eds) (1986), *Mapping the Dynamics of Science and Technology* (London: Macmillan).
Campbell-Johnston, R. (2006), 'If you love Venice, let her die', *The Times* 5 June.
Chirivi, R. (1976), *L'Arsenale di Venezia, Storia e obiettivi di un piano* (Venice: Marsilio).
Chirivi, R. (1983), 'L'Arsenale veneziano – un itinerario monumentale', in R. Chirivi et al. (eds).
Chirivi, R., Gay, F., Crovato, M. and Zanelli, G. (eds) (1983), *L'Arsenale dei Veneziani* (Venice: Filippi).
Coleman, J. (1988), 'Social capital in the creation of human capital', *American Journal of Sociology* 94, 95–121.
Concina, E. (1991), 'La casa dell'Arsenale', in A. Tenenti and U. Tucci (eds).
Conti, S. (1993), 'The network perspective in industrial geography: towards a model', *Geografiska Annaler B* 75:3, 115–30.
Costa, P. (1990), 'Venezia possibile, Venezia probabile, Venezia certa: il turismo', in Fondazione Istituto Gramsci Veneto (ed.).
D'Agostino, R. (2002), 'Piani e progetti per nuovi scenari urbani', in E. Barbiani (ed.).
Dematteis, G. (1993), 'Sistemi territoriali locali come nodi di reti', in A. Peano (ed.).
Dematteis, G. (1999), 'Cities as nodes of urban networks', in P. Bonavero et al. (eds).
Dematteis, G. and Ferlaino, F. (eds) (2003), *Il mondo ed i luoghi: geografia delle identità e del cambiamento* (Turin: IRES Piemonte).
De Michelis, C. (2005), 'Vent'anni dopo' in F. Sbetti (ed.)
Distefano, G. and Paladini, G. (eds) (1997), *Storia di Venezia 1787–1997. Dalla Monarchia alla Repubblica* (Venice: Supernova-Grafiche Biesse).

Dorigo, W. (1973), *Una Legge contro Venezia. Natura, storia, interessi nella questione della città e della laguna* (Rome: Officina).

Dri, M.G. (1993), 'La società veneziana', in M. Reberschak (ed.).

Edwards, B., Foley, M.W. and Diani, M. (eds) (2001), *Beyond Tocqueville: Civil Society and the Social Capital Debate in Comparative Perspective* (Hanover: University Press of New England).

Emanuel, C. and Governa, F. (1999), 'The local milieu', in P. Bonavero et al. (eds).

Fabbri, G. (2005), *Venezia: quale modernità. Idee per una città capitale* (Milano: Franco Angeli).

Fine, B. (2002), 'They f**k you up those social capitalists', *Antipode* 34:4, 796–9.

Foley, M., Edwards, B. and Diani, M. (2001), 'Social capital reconsidered', in B. Edwards et al. (eds).

Fondazione Istituto Gramsci Veneto (ed.) (1990), *'Idea di Venezia', Atti del convegno L'Idea di Venezia, 17–18 giugno 1988* (Venice: Arsenale).

Fontana, G.L. (2002), 'L'economia', in S. Woolf and M. Isnenghi (eds).

Gasparri, S., Levi, G. and Moro, P. (eds) (1997), *Venezia. Itinerari per la storia* (Bologna: Il Mulino).

Giaccaria, P. (1999), *Competitività e sviluppo locale* (Milan: FrancoAngeli).

Ginsborg, P. (1978), *Daniele Manin e la Rivoluzione Veneziana* (Milan: Feltrinelli).

Governa F. (1997), *Il milieu urbano. L'identità territoriale nei processi di sviluppo* (Milano: Franco Angeli).

Governa, F. and Salone, C. (2005), 'Italy and European spatial policies: Polycentrism, urban networks and local innovation practices', *European Planning Studies* 13:2, 265–83.

Grandesso, A. (1988), *Arsenale: quale destinazione d'uso?* Laurea Magistralis thesis, IUAV, (Venice).

Grandesso, A. (2006), Interview with the author 12 October 2006, Venice (cassette recording in possession of author).

Grubb, J.S. (1986), 'When myths lose power: Four decades of Venetian historiography', *The Journal of Modern History* 58:1, 43–94.

Harvey, D. (1985), *The Urbanization of Capital: Studies in the History and Theory of Capitalist Urbanization* (Baltimore, MD: Johns Hopkins University Press).

Indovina, F. (1995), 'Centro Thetis', in A. Bonomi (ed.).

Indovina, F. (2005), Interview with the author 19 December 2005, Venice (cassette recording in possession of author).

Isin, E. (2002), *Being Political: Genealogies of Citizenship* (Minneapolis: University of Minnesota Press).

Isnenghi, M. (1997), 'Fine della Storia?', in S. Gasparri et al. (eds).

Isnenghi, M. (ed.) (2002), *Storia di Venezia: il Novecento* (Rome: Istituto della Enciclopedia Italiana 'Giovanni Treccani').

Knorr Cetina, K. and Cicourel, A. (eds) (1981), *Advances in Social Theory and Methodology: Toward an Integration of Micro and Macro Sociologies* (London: Routledge and Kegan Paul).
La Nuova Venezia (2006), 'D'Agostino: le aree subito a noi', 4 March.
Lalli, D. (2006), Interview with the author 13 October 2006, Venice (cassette recording in possession of author).
Latour, B. (1986), 'The Powers of Association', in J. Law (ed.).
Latour, B. (1992), 'Where are the missing masses? The sociology of a few mundane artifacts', in W.E. Bijker and J. Law (eds).
Latour, B. (1993), 'Ethnography of a "high-tech" case. About Aramis', in P. Lemonnier (ed.).
Latour, B. (1996), 'On actor-network theory: A few clarifications', *Soziale Welt* 47, 369–81.
Latour, B. (2005), *Reassembling the Social: An Introduction to Actor-Network-Theory* (Oxford and New York: Oxford University Press).
Law, J. (1994), *Organizing Modernity* (Oxford: Blackwell).
Law, J. (2001), 'Notes on the theory of the actor network: Ordering, strategy and heterogeneity', published by the Centre for Science Studies, Lancaster University, <http://www.comp.lancs.ac.uk/sociology/papers/Law-Notes-on-ANT.pdf>, accessed 18 January 2008.
Law, J. (ed.) (1986), *Power, Action and Belief* (London: Routledge and Kegan Paul).
Law, J. and Mol, A. (1995), 'Notes on Materiality and Sociality', *Sociological Review* 43, 274–94.
Lefebvre, H. (1991), *The Production of Space* (Oxford: Blackwell).
Lemonnier, P. (1993), *Technological Choices. Transformations in Material Cultures since the Neolithic* (London and New York: Routledge).
Lombardi, G. and Paternò, R. (1992), *Riqualificazione dell'Arsenale – Il quadro urbanistico* (Venice: Fondazione Venezia 2000).
Lugato, D. and Ruol, M. (eds) (1994), *L'arsenale di Venezia. Una metodologia per la conservazione* (Milan: Città Studi Edizioni).
Mazzariol, G. (1985), 'Tre progetti per Venezia rifiutati: Wright, Le Corbusier, Kahn', in G. Romanelli and L. Puppi (eds).
McKie, R. (2006), 'Send for Disney to save Venice', *The Observer* 4 June.
Morand, P. (2002), *Venices* (London: Pushkin).
Musu, I. (ed.) (1998), *Venezia sostenibile: suggestioni dal futuro* (Bologna: Il Mulino).
Paladini, G. (2002), *Uscire dall'isola. Venezia, risparmio privato e pubblica utilità. 1822–2002* (Bari: Laterza).
Paruzzolo, A. (2006), Interview with the author 12 January 2006, Venice (cassette recording in possession of author).
Paruzzolo et al. (eds) (2007), *Thetis all'Arsenale 1997–2007* (Venice: Grafiche Veneziane).

Peano, A. (ed.) (1993), *Insegnamento, ricerca e pratica urbanistica* (Turin: Cortina).

Perulli, P. (ed.) (1989), *Città della Scienza e della Tecnologia* (Venice: Arsenale).

Pes, L. (2002), 'Gli ultimi quarant'anni', in M. Isnenghi (ed.).

Piva, M.G. (2003), 'Un approccio multidisciplinare agli interventi di salvaguardia', *Quaderni Trimestrali del Consorzio Venezia Nuova* 3/4, 1–10.

Putnam, R. (1993), *Making Democracy Work: Civic Traditions in Modern Italy* (Princeton, NJ: Princeton University Press).

Reberschak, M. (ed.) (1993), *Venezia nel secondo dopoguerra* (Padova: Il Poligrafo).

Richardson, J. (ed.) (1986), *The Handbook of Theory: Research for the Sociology of Education* (New York: Greenwood Press).

Romanelli, G. and Puppi, L. (eds) (1985), *Le Venezie possibili* (Milano: Electa).

Rullani, E. and Micelli, S. (1998), 'La produzione immateriale a Venezia: verso un'economia del post-fordismo', in I. Musu (ed.).

Ruskin, J. (1985), *The Stones of Venice* (New York: Da Capo Press).

Sbetti, F. (ed.) (2005), *Venezia 2000. Cultura e impresa* (Venice: Fondazione Venezia 2000).

Tenenti, A. and Tucci, U. (eds) (1991), *Storia di Venezia: Dalle origini alla caduta della Serenissima, Temi: Il Mare* (Rome: Istituto della Enciclopedia Italiana 'Giovanni Treccani').

Turco, A. (1988), *Verso una teoria geografica della complessità* (Milan: Unicopli).

Woolf, S. and Isnenghi, M. (eds) (2002), *Storia di Venezia: l'Ottocento e il Novecento. Il Novecento* (Rome: Istituto della Enciclopedia Italiana Giovanni Treccani).

Zanelli, G. (1983), 'L'Arsenale – Storia di industria e città', in R. Chirivi et al. (eds).

Zanetto, G., Antonelli, C., Bagnasco, A., Dematteis, G., Ferlaino, F., Roscelli, R. and Salvatori, F. (2003), 'Il territorio come produttore di conoscenze', in G. Dematteis and F. Ferlaino (eds).

Zucconi, G. (2002a), 'L'Istituto Universitario di Architettura (IUAV)', in M. Isnenghi (ed.).

Zucconi, G. (2002b), *La grande Venezia. Una metropoli incompiuta tra Otto e Novecento* (Venice: Marsilio).

Chapter 10
New and Old Associations as Trusting Networks: Tracing Social Capital in Jyväskylä

Martti Siisiäinen

Voluntary Associations as Social Capital

The relationship between the state and social capital (in its various forms) is complex and dialectical. With research on social capital rapidly increasing since the mid-nineties however, it seems that some of the disputes have been settled. At present, more and more researchers working on questions of social capital and trust have, for example, reached the conclusion that there is no convincing evidence of a positive correlation between acting in associations (on an individual level) and the production of trust as such. It has however been shown that there is a strong correlation (at least in Western societies) between the number of (democratic) associations and the average number of citizens' memberships on the one hand, and the level of trust on the other (see Strømsnes and Wollebæk 2006; Trägårdh 2007; Kankainen 2007). Such findings suggest the importance of further study into the influence of voluntary associations on local communities and national polity. This chapter thus focuses on voluntary associations as a means of acting collectively and as a way of (re)producing social capital via cooperation between members. In comparison with other types of social networks, voluntary associations can therefore be understood as organized social capital. Significantly, the establishment of an association formalizes already existing networks, ties participation to membership roles, stabilizes the 'organization of trust' and, to a large extent, makes it autonomous from individual members. This process by which associations attain such a level of autonomy therefore increases the possibilities for them to regulate their relationships with their own environment (see Luhmann 1988).

Although the approach adopted in this chapter does not emphasize the trust component in social capital as understood by Robert Putnam (for criticism, see Cannone in this volume), this concept can still be related to associations in several ways. First, the interactions which lead to the formation of an association require a minimum level of trust between potential founding members. Second, the continuity and effectiveness of a modern voluntary association necessitates that members rely on their association to work for the common interests and/or goals of the members (see Simmel 1968 (1908), 277; Richter 1985, 240). Third, successful interactions between different voluntary associations, or between associations and the state or municipal institutions, are also based on trust between all partners.

There are however, other ways for various types of actors to establish their reliability and trustworthiness in different contexts. Multiple legitimization procedures are available for this purpose. Collective movements and other less formally organized groups, for instance, can present themselves as collaborative and law-abiding endeavours by adopting an official bureaucratic form – the form of a 'registered association' in the case of Finland – thereby convincing their partners to take them more seriously (see Siisiäinen 2007a).

Finland has a unique registry of associations containing data on all new voluntary associations (and thereby organized interests) registered since 1919 (when the *Law on Associations* was issued). When associations are registered, they become legal subjects with the right to plead a case in court and to own property. Such a registration also absolves the group's leaders who become board members of personal responsibility for the deeds of the association. It is typical in Finland for all major movements to register as official associations when they become established as national movements. Even the Communist movement, which was illegal in the 1920s and 1930s, tried to establish registered associations under various disguises (see Siisiäinen 1990). Significantly, the act of registration in Finland increases an association's symbolic capital by guaranteeing the organization a degree of prestige that comes with the acronym *ry*, which stands for 'registered association' (*rekisteröity yhdistys*). The registration of an association means, firstly, that it more or less accepts the principles of a formal organization, thereby rejecting 'anarchic' or disorganized forms of collective action. Secondly, registration gives a legitimate label to any association. This 'prestige' function, which is associated with highly esteemed voluntary organizations, is a central way of developing associational symbolic capital (see Siisiäinen 1988; 2003a).

Since voluntary associations have been established as 'trusting networks', they can be mobilized in different social fields as social capital, as resources based on the strength of the membership group and as networks that can be mobilized with the assistance of the association. The notion of a voluntary association as social capital can therefore be understood as a collection of mobilizable resources, produced by the association for the realization of its specific interests. After all, the formation of an association not only creates solidarity and trust among its members, but also institutionalizes and gives a name to the 'capital' that is thereby being accumulated. Social capital accumulated in voluntary associations can, to some extent at least, be allocated to the leadership of the association, with some part of it filtering down to rank-and-file members (see Bourdieu 1980, 3; 1986, 251; Siisiäinen 2003a). As with individuals, however, the amount of associational social capital is equivalent to the amount of economic, social and cultural capital that the association is able to mobilize through those actors (individual and collective) that it has managed to tie into its networks (Bourdieu 1980).

Registered associations are then places where symbolic capital is produced and consumed, because their influence on society and their partners is based on mutual cognition and recognition. Indeed, this is why Bourdieu uses the concept of legitimate capital as synonymous with symbolic capital (see Bourdieu 1980, 3; Bourdieu 1986). Voluntary associations also take part in struggles over the (re)production of social groups and classes: voluntary associations, as actors in the making of new collective identities, often become nodal points of group formation (see Laclau and Mouffe 1985; Bourdieu 1985; Siisiäinen 1986). According to an anti-essentialist view, group formation does not have privileged nodal points, and this is why social movements and voluntary associations can be seen as precursors to future group structures.

Names play a central role in the formation of associational identities, both as signs and as symbols (see Bourdieu 1980; Siisiäinen 1988; 1992). The naming of an association makes it distinct from other associations – whilst also grouping it together with similar associations. The act of naming both individualizes and helps to build an associational collective. Naming a group defines both the group itself and its difference from others, or in some instances, their mutual similarity (see Palonen 1990; Pekonen 1993, 10–18). The goal or objective of the association, as understood by its members, describes its relationship with the surrounding society. The name of the association then acts as a public sign and/or symbol and thus as the apparent inducement (and filter) to the membership in the association. Voluntary associations have consequently become some of the most important agents in transforming potential groups into actual groups through processes of naming and other forms of representation (see Bourdieu 1980; Siisiäinen 2007b).

Created through dialectical interactions between the founding members and the social environment (or fields), associations do not appear from nothing, and yet they are not produced by deterministic mechanisms. The mere fact that associations are born out of communication that distinguishes them from their own context means that relationships of dependence may exist between associations and their own 'turf' preconditions (Luhmann 1984; Siisiäinen 2003c). Such connections become especially apparent when associations founded in the last ten years are compared to those created in preceding decades. These differences cannot easily be explained, but among the trends that have influenced associational changes on the societal level are normally identified: globalization, individualization, increased division of labour, the blurring of social classes and interest structures, and a more general cultural change (see Siisiäinen 2003b). The contrast between older and newer associations is a central theme in the following analysis of associational development in one specific Finnish urban context.

Indeed, this chapter examines the role of associational social capital as embodied in the field of welfare policy in Jyväskylä, a city where different kinds of relations between associations and the various institutions of local government play an important role. I begin with a discussion of associations as social and symbolic capital; moving on to consider the relationships between local forms of national government and voluntary associations, within a context of a European theoretical typology of polity. This chapter also analyses the formation and characteristics of registered associations in Jyväskylä, and the inner differentiation of the system of associations since 1919. Particular attention is paid to the interactions between associations, and between these later and different organizations and institutions in place. The chapter concludes by discussing the role of voluntary associations in the political and welfare system – or local political field – of the city of Jyväskylä, and examines more generally the changes taking place within the system of voluntary associations.

Jyväskylä is currently a mid-sized Finnish city of approximately 128,000 inhabitants (see Figure 5.1, p. 122). Officially founded in 1837, Jyväskylä was one of the main centres of the nineteenth-century Finnish nationalist movement, as well as a transport hub and educational and cultural centre for the surrounding province of Central Finland. During the twentieth century the city's development was characterized by two significant types of progress: the growth (at the Finnish scale) of major paper and metal industries, and the ascendancy of its educational and cultural institutions. Jyväskylä's paper industry dates back to 1872 and its metal industry from the 1920s. The development of these industries arguably polarized the structure of workplaces in the city, resulting in the creation of large factories on the one hand, and very small family enterprises, on the other, with very few mid-sized enterprises in between.

The cultural field of Jyväskylä has also been dominated by prominent educational institutions. During the nineteenth century these consisted primarily of the first Finnish-speaking secondary school (1857) and a regional college for teacher training (1863). Since then, numerous schools and educational institutions have helped shape the city's cultural field, as well as influencing its occupational structure. The image, and more broadly, the development of Jyväskylä, has been marked by the shift towards educational institutions, especially since the teacher training college became a university in the mid-sixties. Since then, the Jyväskylä region has become one of the few growing areas in Finland, driven most visibly by the presence of the university and certain spearhead enterprises, such as Metso Industries' paper machine manufacturing (Nummela 1997). Cultural and associational life in the city is therefore now largely dominated by various groups working within educational institutions (the university, secondary schools and polytechnics), encompassing both staff and students. Since cultural institutions play such a central role, and since a significant proportion of the population is consequently made up of young well-educated people, Jyväskylä offers a good case for studying broader patterns of change amongst voluntary associations. People belonging to this group have always taken a key role not only in creating associational innovations but also in spreading them from place to place (see Nummela 1997; Siisiäinen 2002a).

Associations and Local Fields

Associations as social capital can have different kinds of impacts on the social system, both at a national and a local level. Football clubs and choirs may indeed serve as consensus-building forces in many societies, but this function can be interpreted sociologically in a variety of ways. Putnam understands it as a kind of 'rosy circle', in which clubs or associations create trust and are themselves created out of trust (see Putnam 1993; 1995; Siisiäinen 2003a; also, Cannone in this volume). Max Weber's ideas about the function of choirs, as presented in his speech at the first meeting of German sociologists in 1910, however, offer an interesting contrast to Putnam's view. According to Weber (1911) one of the main functions of associations, such as singing clubs, is their capacity to educate, or fashion, 'good state citizens', by directing people away from activities that might cause political unrest and trouble for the ruling elites, and by integrating their members into the political system through the relations of domination (*Herrshaftsverhältnisse*) that exist within such associations. The principal way of maintaining the internal division of labour and the organizational basis of the associational elite was to adopt a hierarchic bureaucratic model as the dominant form of government (Weber 1911; Siisiäinen 1986).

Football clubs and choirs however, have at times also been involved in instigating protests and even revolutions. It can be argued then that the capability of a society to deal with or exploit conflicts is at least as important to social stability as its mechanisms for fostering forms of non-conflictual social capital (see Siisiäinen 2003a). In sociological analyses, this means that the functions and societal impacts of associations cannot not be taken for granted, but have to be studied in their contexts, that is, in relation to the social fields where they are active (or 'struggling').

One of Bourdieu's main contributions to the sociology of associations is the concept of the 'social field'. It is here that the meaning and the current value of the concept of social capital is tested. In Bourdieu's words:

> In analytic terms, a field must be defined as a network, or a configuration, of objective relations between positions. These positions are objectively defined, in their existence and in the determinations they impose upon their occupants, agents or institutions, by their present and potential situation (situs) in the structure of the distribution of species of power (or capital) whose possession commands access to the specific profits that are at stake in the field, as well as by their objective relation to other positions (domination, subordination, homology, etc.). (Bourdieu and Wacquant 1992, 97–8)

Fields develop from relational interactions between agents who have different objectives but contested positions in the space, and who are endowed with different amounts of capitals, which are valuable in that specific field. It is important to note that the total agency in social fields consists of different types of actors such as individuals, social movements and associations, organizations and institutions. Voluntary associations are the most important ways for individual actors to increase their influence by joining forces together. This has been the main reason for mass mobilization throughout time, and even today it is the main reason for the formation of associations for various purposes.

The borders of social fields, however, are always contested and subject to historical development, and therefore it is not easy to define them or capture their components in any detail (see Bourdieu and Wacquant 1992; Alanen 2007). It would be thus very difficult to adapt Bourdieu's concept and criteria of the field strictly to a local community. In this analysis, I use the concept of the 'field' simply as a general methodological means guiding the understanding of the functioning of local societies. Towns and municipalities consist of different, relatively autonomous, fields. In these fields, actors – individual and collective – struggle over stakes that have been defined or negotiated as valuable (as a combination of valuable elements of different forms of economic, cultural and social capital). The main point here is to conceive of local society as a totality of dynamic spaces. Taken synchronically, local fields can be understood as organized systems of social reproduction, since their relative stability is dependent on the continuity of more or less institutionalized relations between actors occupying relevant positions in the space in question. From a diachronic point of view, fields are conflict-laden spaces, contested by actors trying to gain their own stakes, or more favourable positions.

The primary local fields in Jyväskylä that can be approached using the available data are the political field, including local welfare state arenas, and the field of 'cultural production' (see Bourdieu 1992; 1996). In concrete terms these two fields cut across each other in many ways. The focus here is on the political field of welfare, which is the most important arena for local associations. Voluntary associations are understood as inter-actors investing varying amounts of their capitals in political or cultural fields. However, the proportion of associations devoting their actions primarily and consciously to a certain field, for example, politics is relatively small, such that they are not always clearly visible in the urban public sphere. On the other hand, associations taking part in the field of cultural production often have a more wide-reaching connection to other fields, typically contributing to the functioning of these other fields through the unintended consequences of their actions (Kangas 2003). In Finland, the close connections between local political government and associations that have developed in local fields is to a large extent taken for granted, which is why they are not often made visible.

Civil Society and the State: Contextualizing the Finnish Case

Recent studies on social and political participation, association memberships and social capital (including trust) have shown that the Nordic countries have important common characteristics, which distinguish them from most other countries. In contrast to what some researchers have written about the presumed disastrous effects of close state–civil society relations on critical associational life, the public sector and state institutions have played a central role in the creation of a high level of social trust in the Nordic countries, as showed by several comparative studies. As international scholarship has produced more fine-grained data, it has become clear that there is no convincing evidence for the hypothesis that strong, supportive welfare states would 'crowd out' civil society (see Curtis et al. 2001; Rothstein 2001; Trägårdh 2007; Strømsnes and Wollebæk 2006; Warren 1999; Wallace 2004; Baer 2007; Morales and Geurts 2007). On the contrary, it has become evident that strong, easily approachable, democratic associations connected with an equally strong, open, inclusive, corporate state are among the most favourable contexts for the development of a high level of social capital and trust (see Schofer and Fourcade-Gourinchas 2001).

In the 1970s and 1980s Nordic welfare states created the foundations for mass solidarity by uniting different classes and strata into a common collective – the nation state – whose members were connected to each other by principles of citizen obligations and reciprocity based on compromises of interest. A shared identification and the abundance of welfare benefits made possible the creation of a feeling of sameness and thereby confidence in the notion that people coming from different spheres or strata of society, including the middle classes, were mutually dependent, were 'rowing in the same boat', and therefore had an interest in participating (see Crow 2002, 41–2). These factors are arguably the most important basis for the comparatively high degree of generalized trust amongst citizens in the Nordic countries (see Warren 1999; Stolle and Rothstein 2002; Wallace 2005; Trägårdh 2007).

On the basis of earlier studies (see Esping-Andersen 1990) Eva Schofer and Marion Fourcade-Gourinchas have developed a synthesizing typology for state - civil society relationships, which helps to analyse international differences in association activities. The typology was designed for comparisons at the national level but at least from the Nordic perspective its core arguments hold true on a local level as well.

The typology makes distinctions, first, between *statist* and *non-statist* (sometimes called 'liberal') societies (the main criterion: freedom and pluralism of associational development); and second, between *corporatist* and *non-corporatist* societies (the main criterion: tripartite bargaining between unions, the private sector, and government). By cross-tabulating these two factors or dimensions, four types or regimes can be distinguished. The Nordic countries, together with the Netherlands, can most clearly be characterized as countries with close (neo-corporatist) mutual relations of associations with the state (and local government). In this relationship the state/local government is both the target of reform proposals and demands, and a potential supporter of association initiatives, and thereby, on the second level, an inclusive integrator of associations. At the same time, Nordic states guarantee a relatively large democratic and free space for associations to act in ways of their own choice, both on the national and the local level. In this vein, associations in Finland have been understood as legitimate articulators of individuals' aspirations, empowering individuals chiefly as members of broader collectives. Individual citizens are thought to interact with the political system in the role of association members, not as individuals. The legitimacy of the state was derived from its function as the representation of civil society (Schofer and Fourcade-Gourinchas 2001, 811–13).

In an ideal situation, Nordic citizens would be aware of the existence of a relatively well-functioning system of associations that is capable of mediating and representing their interests before the political and administrative institutions. The borderlines between civil society and the public sphere have been thus often blurred, especially on the local level. In local government, associational governance has sometimes even been mixed up with public affairs, with the same people active in both sectors (see Liikanen 1996; Stenius 1987; Trägårdh 2007).

The specific features in the political, welfare and ideological history of the Nordic countries have also left their marks on Nordic ways of conceptualizing and studying social capital. Starting from the early expression of interest in the early nineties, the Nordic understanding of the concept of social capital has often taken ideas from neo-corporatist theorizations on associations' relations with the state, and has drawn on Bourdieu's conceptualization, including political subcultures, folk organizations and social movements (see Ruuskanen 2001; Siisiäinen 2003a; Trägårdh 2007).

Finland shares all of the central characteristics of a non-statist corporate type of polity: a relatively strong and open state with an extensive and efficiently functioning system of associations. The state and its municipalities have been, and in many ways still are, intertwined as providers of welfare, education and cultural services. As Schofer and Fourcade-Gourinchas's research shows, Finland, together with the other Nordic countries, has the highest number of association memberships voluntary associations per capita in the world (see Morales 2001; Baer 2007; Siisiäinen 2007b). International comparisons also reveal that Finns, in comparison to citizens of other countries, are some of the most likely to express their social capital in formal voluntary associations (Siisiäinen 2007b). The exceptional fact about Finns is that they have developed fewer alternatives to associational action (informal networks, social life with workmates, relatives or neighbours) than people in most countries (Wallace 2005; Baer 2007). On the other hand, Finns are more active than other EU-citizens on the average as users of the Internet and mobile phones (see World Bank 2005). Aside from the exceptional decades between the two World Wars, the tradition of close relations between the state and civil society has been maintained from the beginning of the twentieth century to the present day, with the institution of registered associations playing a key role in its continuation.

Registration of associations in Jyväskylä, 1920–2002

While the history of voluntary associations in Finland dates back to the beginning of the nineteenth century, the mass base for the modern system of associations was created at the end of the nineteenth century and during the first decade of the twentieth century. Since the 1919 Law on Associations more than 170,000 new associations have been registered in Finland (see Stenius 1987; Alapuro et al. 1987; Siisiäinen 1990; 1992). From the 1920s to the present day, a total of 2,500 associations have been registered in the town of Jyväskylä. As is the case in Finland in general, there are no signs of a general crisis of organized social capital in Jyväskylä (Siisiäinen 1999; 2007b; Baer 2007). On the contrary, more new associations have appeared during the last ten years than ever before, in such a short time. During the record-breaking year of 1997 almost twice as many new associations were registered as during the earlier record years in the 1970s and 1980s (see Figure 10.1).

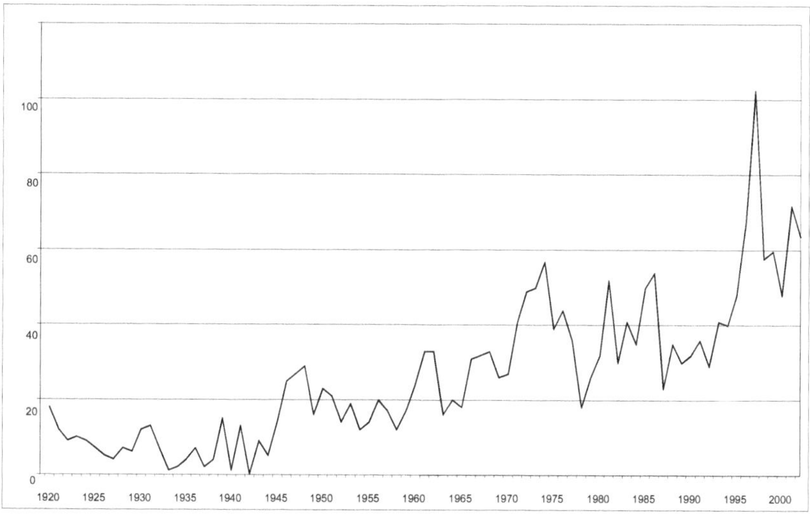

Figure 10.1 Registration of voluntary associations in Jyväskylä 1920–2002

Source: Siisiäinen 2002a, 19.

In addition to the fact that the establishment of new associations in Jyväskylä has been accelerating, it is important to take into consideration, for example, what kind of associations or what types of civil participation are declining or flourishing, and what impact these associations are having in their local fields. Table 10.1 helps us to identify changes in the types of associational activities taking place during the past decades. The table shows how the accumulation of registrations has been continuous, especially since World War II, such that the dominance of the formal registered type of voluntary association presents as an ongoing phenomenon. The most recent rise in the volume of registrations coincided with Finland's recovery from the deep recession of the early 1990s. It is characteristic of this period that the flood of new associations was not given much impetus by active protest movements as was the case during many other active periods (for instance 1944–1948; 1968–1976; the turn of the 1980s – see Siisiäinen 1990; 1992).

The number of economic/professional and party-political associations has dropped dramatically since the 1970s, whereas the proportion of recreational associations has been growing: more than 60 per cent of the new associations established since the mid-1990s are cultural associations, sports clubs or other hobby associations. At the same time the majority of new associations have been created to advance very specific and specialized interests (see Negt and Kluge 1976). However, there is evidence that the main trend has been that of moving away from generalist multi-task associations towards the creation of specific single-task organizations. The local political field consequently seems to be dominated by a multitude of specific associations only very indirectly interested in local formal politics or in interest groups' activities.

At first sight it might seem that the latest wave of associations has produced exactly the kinds of associations that are so dear to the Putnamian ideas of social capital (choirs, sports clubs, general hobby groups, etc.). The development of specific associations, however, indicates that people invest only a small part of their time and energy in new associations, thereby minimizing the risk of 'wrong investment' connected with the membership. It has been typical of the older bonding associations (Putnam 1993; Cannone in this volume) that they have acted as arenas for the co-production of various common goods and as places of relatively democratic discussion or dispute. By succeeding in bringing together people from different social or political origins to discuss and interact, they have promoted the development of solidarity and generalized trust (bridging social capital).

The new associations, whose internal functional model has much in common with the logic of consumption, do not seem to be very appropriate for the creation of either bonding social capital among members or solidarity between various social groups. Commitment in general requires tying individuals or personalities to a social system, to its substructures and values. People can be said to be committed if they are willing to offer their energy, to stay loyal and attach themselves to the object of their commitment instrumentally, morally or emotionally (see Kanter 1968). The concept of social ties describes connections between actors or collectives, whereas solidarity has to do with the forces which – in the form of loyalty or moral obligations – join the collective and the individual together (see Crow 2000; Touraine 2000). Thus, it seems that the associational trend is moving towards specialized cultural and recreational associations, but

Table 10.1 Registration of different types of voluntary associations in Jyväskylä 1920–2001

Type of associations	1920–29		1930–39		1940–49		1950–59		1960–69		1970–79		1980–89		1990–94		1995–2001		Total	
	N	%	N	%	N	%	N	%	N	%	N	%	N	%	N	%	N	%	N	%
Political	15	17	6	9	23	17	22	13	29	11	106	27	48	13	13	7	6	1	268	
Economic and professional	25	29	30	45	42	30	54	32	89	34	116	30	83	22	24	14	73	16	536	
Social welfare	6	7	3	5	10	7	18	11	21	8	27	7	34	9	26	15	45	10	190	
Culture and education	8	9	8	12	19	14	30	18	42	16	46	12	75	20	32	18	111	24	371	
Sports	8	9	11	16	15	11	13	8	17	6	24	6	54	14	35	20	101	22	278	
Other hobbies	11	13	3	5	11	8	16	10	38	14	48	12	45	12	27	15	78	17	277	
Religion	6	7	2	3	–	–	2	1	4	2	5	1	5	1	8	5	15	3	47	
War and peace	2	2	2	3	7	5	5	3	11	4	1	0.3	7	2	–	–	5	1	40	
International	–	–	–	–	4	3	5	3	2	1	2	1	11	3	5	3	7	2	36	
Other	6	7	2	3	8	6	4	2	13	5	12	3	20	5	8	5	16	4	89	
Total	**87**	**100**	**67**	**100**	**139**	**100**	**169**	**100**	**266**	**100**	**387**	**100**	**382**	**100**	**178**	**100**	**457**	**100**	**2132**	**100**

Source: Siisiäinen 2002, 20.

as yet we lack convincing sociological evidence of their ability to create the Putnamian sort of social capital in terms of generalized trust.

The Nordic cases show that generalized trust has been prospering especially in contexts where associations have been able to connect people from different classes with more or less collective ties (Rothstein and Stolle 2002; Trägårdh 2007). In Finland, the proportion of 'consumption-like' associations has increased during the last 20 years, whereas many associations based on a specific ideology and formally political associations have declined to a crisis point, as both membership figures have fallen and young people have shunned their activities.

Associations as Networks of Social Capital in Jyväskylä

Membership and Economic Resources

The data used in this chapter was collected as part of a survey on active voluntary associations in Jyväskylä. The estimated number of genuinely active associations in the town is around 800–900. Nearly 350 associations returned the questionnaire. Local political party and trade union branches are underrepresented among those that returned the questionnaire. However, in general the data is representative of associations that are locally active and want their opinions to be heard and included in the overall picture provided by the research.

The survey was addressed to associations and therefore the respondents represent their associations' perspectives, not their personal opinions. The respondents belong to the operative leadership of associations, selected through the procedures of representative associational democracy following the statutes of the Finnish Law on Associations. (Re)elected annually, they can be said to represent more or less the general opinions of the membership. Perhaps the motivation of the leadership in typical associations can be characterized by a combinations of Max Weber's ideas of the concentration of power in the hands of those who have most skills, and the results of Pamela Oliver's studies, which suggest that leadership requires a sense of responsibility and to some extent willingness to sacrifice one's time and resources without any certainty of success – 'If you do not do it, nobody else will' (Weber 1911; Oliver 1984).

The majority of the respondents were association chairs (64 per cent), 10 per cent were salaried managers or other staff, 10 per cent secretaries, and the remaining 16 per cent other activists. In terms of socio-economic status, one third of the respondents were upper level employees, 31 per cent represented lower level employees, 9 per cent students, 5 per cent workers and 15 per cent pensioners. The male/female ratio among the respondents was quite even (51–49), although the chairs mostly consisted of men, whilst among the secretaries, women were clearly in the majority (Siisiäinen 2002a, 27–9).

Most of the associations are economically 'poor': about 60 per cent of those surveyed have a yearly budget of less than €10,000. Therefore, the principal resource of these associations is their members. As the survey focused especially on associations, no detailed information on the individual members themselves was collected. Local associations in Jyväskylä are not very large: the average number of members was 327 and the median, 75. Associations are nevertheless somewhat bigger than in a comparable Norwegian town (see Wollebæk et al. 2008). Among the associations, there are a few big trade unions whose branches have several hundreds, even some thousands of members. The problem with the local branches of central unions is that, while the central unions may have a lot of social and even economic capital, local branches and their memberships (especially in trade unions and political parties) are typically passive and subjugated to the centre.

The average number of active working members is 76, that is one fourth of those who have paid their membership fees. The overall number of members increased in half of the associations from 1994 to 2001. In one third of the associations the number of regularly working membership has increased. The average number of volunteers working weekly in associations is 17, which is clearly above the national average (11 per association). The number of volunteers has also been continuously rising since the mid-1990s (Pyykkönen 2002, 97–103). There are no data on the gender division of the membership but the association leadership is divided evenly: 53 per cent of chairs and 52 per cent of secretaries are men.

Locality and the Level of (Regional) Activities

The first nationwide mass mobilization at the end of the nineteenth century united the country with civic organizations ranging from political to hobby associations. These voluntary organizations had their roots in particular localities, but tended to be tied into a nationally integrated (political) culture. During this period a four-level model of organization was established and adopted by all large movements. Grassroot associations (village associations) that were members of a municipal parent organization operated at the bottom. Regional organizations mediated between municipal associations and the national central organization or union (see Stenius 1987; Ursin 1990; Siisiäinen 1991). Some associations and central unions were also members of their international parent organization, and so, to some extent, bound together (for example, international sports organizations, Scouts, etc.). In the long run, modernization meant that the remaining local units were tied together on a national level, thereby making these voluntary groupings 'supra-local' (see Richter 1985, 213–14; Tenbruck 1989).

Membership in a large national central federation or union can empower local associations symbolically, but this kind of support is always conditional, and as a rule, top down. 'Wild' associations have more freedom in this respect, but at the expense of losing support provided by national level organizational capital. In Jyväskylä 73 per cent of associations are members of a central organization. Almost all older associations belong to central organizations in comparison to only 40 per cent of those registered in 1995–2001. Clear differences exist between association types: in practice all political associations belong to central unions, as do four-fifths of economic-professional associations, but less than half of hobby associations. The independence of the newest associations from national central organizations is one indication of the loosening of national ties. It also suggests that many of the new associations are more flexible and more adaptable to social change. Older associational subcultures, in comparison, tended to be conservative and slow to accept and reflect change, as their rigid organizational structures fostered cultural inertia. What this shift might mean to the nature of social capital embodied in these associations is largely an open question to be answered by studies to come.

New associations govern themselves by relatively lax organizational or bureaucratic structures, and sometimes avoid them entirely. Member recruitment is more informal, and the process of decision-making often resembles the model of a direct democracy. Such features could suggest that the role of new associations in the neo-corporate system of governance is probably weakening, as their ability to tie members into their collective interests and ideals is decreasing. On the other hand, the dispersion of social structures, increasing individualization, and the corresponding weakening of associational ties, combined with the trend towards specialization in associations, arguably makes it more difficult to connect various wild associations into sub-cultural politics. In Jyväskylä, strong ideological subcultures seem to have vanished and developed constellations of new associations that are finding their mediators in other spheres of life (sports, international projects, Internet networks, etc.). Associational subcultures are based on the existence of a general common framework overlapping with more specific interests of distinct associations. It seems that new associations find it hard to develop such general frameworks when following the logic of (individualistic) co-consumption instead of (collective) co-production. It is not clear whether, and on what preconditions, individualistic ties to associations may transform into 'real' enduring collective action. In the Jyväskylä study it turned out that new associations have more cooperation with business organizations and are less predisposed to unconventional repertoires of collective action, such as protesting.

Naming of Associations as Symbolic Capital

The general trend towards individualization and specialization over the last two decades is also apparent in the system of associations. The naming practices of voluntary associations are a case in point. Chiefly, there has been a move away from names that represent collective identities and reflect social structures or large societal formations, towards names as 'logos' or signs of habits or lifestyles (on the concepts of sign and symbol, see Opp de Hipt 1987). In a way, the fragmentation of the structure of social forms of interest is reflected in the changing names of the associations.

The first general observation concerning the names of the new associations in Jyväskylä is their 'internationalization'. The proportion of association with English names has increased from 2 per cent in 1960–1979, to 6 per cent in 1980–1994, and up to 12 per cent between 1995 and 2001. The proportion of Swedish names – the second official language in Finland – remained constant at 1 per cent. It is interesting that the proportion of other or unidentified languages and different kinds of acronyms rose to 7 per cent of all names at the turn of the century. On the other hand, in the record-breaking year of 1997 alone, six immigrants' associations were registered with Finnish names. More than one-quarter of the new names contain foreign loan words, including acronyms (Team Scarabee; Ladies' Circle-27, Jyväskylä). The internationalization of the naming is connected with the specialization of associations for specific hobbies, as expressed in specific and exact names. In the field of sports, names also describe the transition from generalist to specific clubs that are specializing in only one sport (see Piirola 1985; Ojanen 1992).

A second characteristic related to the above is a decrease of ambiguous names and of names that have their origins in proverbs or puns, which was one of the traditions started by associations in the nineteenth century. In this respect, new associations have made a clear break from the past. In the registrations of 1997 only a few names belong to this category and many of them refer to new international cultural items: for example, MaMBA Markkinointimestarit ry (Marketing masters, punning the name of a poisonous snake), Hengen Viljelijät HeVi ry (Cultivators of Spirit, abbreviated to a pun on 'heavy') and sports clubs Jyväskylän Kyky ry (Jyväskylä's Talent), Mäki-Matin Mahtiankat ry (Hillbilly-Matt's Mighty Ducks), all utilizing the fact that the words can be used in many senses or that acronyms can have different meanings in spoken language. It might be an indication of the increasing estrangement of the urban culture from nature that natural symbols or names of natural phenomena or names with references to national cultural heritage have become rare among new associations (Piirola 1985). Modern names with humorous shades of meaning are almost totally restricted to sports clubs (Mäki-Matin Mahtiankat, mentioned above, or Kampuksen Dynamo: The Campus Dynamo). Nowadays there are only a few names containing sonic terms, which was a somewhat common practice in naming associations in earlier decades. Traditional names borrowed from national proverbs were also common in the earlier decades, but now seem to have disappeared almost completely.

Earlier, during the processes of the industrialization and urbanization of Finnish society, associations' names often referred directly or indirectly, symbolically or ideologically, to emergent class structures. This is not the case any longer amongst newly founded associations: a mere 10 per cent of their names refer to a section of a social class or a profession ('Cuppers of Finland';

'The Beef Producers Association of Central Finland', etc.). References to large social classes have disappeared entirely. This coincides with the prevailing tendencies towards the dispersion of a clear division and a weakening of ideological organization that reflects class interests and the unravelling of class-based subcultures (Siisiäinen 2003b). Ideology has also almost completely disappeared from the repertoire of new names, with the only exception being one lonely local branch of the Communist Party registered in 1998. Similarly, names referring to positions on the political map (see Palonen 1990) have disappeared, except for one political newcomer, Jyväskylän Keskustaseura (the Jyväskylä Centre Party Association). These trends could be interpreted as symptomatic of a 'mutation society', in which old ideological co-ordinates no longer function and new coordinates are perhaps only just taking shape (Touraine 1988).

As to regional sources for names, association names connected with subcultures at the national level are becoming rare, which reflects a fading away of nation-wide ideological movements. Local names connected with specific places have kept their position among the new associations, reinforcing the conclusion that the importance of local collective identities has been increasing as quickly as the trend towards globalization. In the formation of collective associational identities, the diminishing role of national factors, as opposed to both international and local ones, is seen in all indicators as representative of the changes taking place in the system of associations, including that of name-giving.

Associational Social Capital in the Local Fields of Jyväskylä

In the preceding part of this chapter, the volume of associational capital has been linked to the number of members and volunteers in associations, and the changing nature of voluntary associations is charted by considering their naming practices. This section focuses instead on associations' interactions that utilize and accumulate social capital in the political welfare field – including the production of welfare services and a cultural policy component. This is done by investigating the number and the character of interactive partners in local fields, and by examining the participation of actors in local government (elected officials; city councillors, etc.) within association boards. The idea is that social capital is not something that can be stored, but instead develops via the relationships between heterogeneous actors. The amount and strength of associational social capital, therefore, can only be depicted in actual practices. Its weight and power is constantly tested in various negotiations – or 'struggles' – in local fields; it is a relational phenomenon requiring both the existence of social networks connecting actors, and the cognition and mutual recognition of their positions.

In Finland, top posts in voluntary associations have traditionally served as rungs on the political ladder, both in municipalities and in parliamentary politics. Traditionally, positions won in local associations have served as the first phase in the development of a national political career. Structural changes in the national public sphere however, have tended to devalue the earlier importance of some institutions, including some voluntary associations, in the function of the political system. Nowadays, success in Parliamentary Elections tends to depend more on the candidates' media image – in addition to the size of her/his campaign budget – than on their accomplishments in the voluntary sector. On the municipal level, however, associations still seem to play a central role, thus continuing the tradition that started together with the development of formal municipal democracy in Finland (see Luomala and Pyykkönen 2002, 136).

In terms of interaction, the most important partners for associations are other associations (79 per cent), followed by the media and municipal officials (49 per cent) and then corporations (47 per cent) (see Table 10.2). There are no significant differences between the old and the new associations in this regard. However, perhaps because they more often belong to associational subcultures, older associations tended to interact more with other associations, whereas newer ones engage more significantly with private companies. It seems that new associations are more often free from ideological or political commitments, which previously might have prevented them from co-operating with enterprises. The 'proximity' of city authorities to associations is also reflected in the fact that 14 per cent of all associations say that they have leading city officials on their boards, 21 per cent report that they have council members, and 33 per cent that they have members of municipal sub-committees or trustees on their boards. The division of local government into sub-sectors (culture, social welfare and health, sports, education, etc.) seems to function effectively in a neo-corporatist sense: local associations mediate much of their communication with the local government via the committees in their own sector.

In the late 1990s, voluntary associations took a central role in the welfare strategies of Jyväskylä, that proclaimed itself to be a city whose economy was based on the presence of advanced services (and involving the third sector). A House of Partnership was founded in Jyväskylä as a joint-venture between the city and the Central Federation of Social Welfare and Health in 2000, providing premises for almost 30 associations. The city has also been active in establishing a foundation (TEKEVÄ) operating as a social enterprise, one of only few of its kind in the whole province (see Möttönen and Niemelä 2005).

Returning to the survey undertaken, associations were also asked to respond to a question about the importance of their cooperation with the public sector/city and companies. The orientation of different types of associations can be summed up in the following typology (Luomala and Pyykkönen 2002, 144 – see Table 10.3). Sports associations, the closest prototype of the new type of associations, interact with both companies and the city (including officials, local politicians and elected officials, as well as the sector administration), whereas

Table 10.2 The year of registration and the frequency of associations interacting with different partners (proportion of 'yes' answers)

Interaction with organization	Year of registration		
	1919–89	1990–2001	All associations
	%	%	%
City authorities	49	50	49
Elected officials	29	25	28
Companies	44	52	47
Other associations	84	72	79
Citizens	28	31	29
Media	63	59	61
Researches	19	18	19
Total	**N = 171**	**N = 105**	**N = 276**

religious and international associations do not regard either of these kinds of activities as especially important. And, perhaps unsurprisingly, economic and professional associations regard companies as important partners, whereas welfare, political associations consider public sector as an important direction for interaction.

Table 10.3 The importance of cooperation with public sector/the town and private sector in different types of associations

		The importance of cooperation with companies	
		High	**Low**
The importance of cooperation with public sector	**High**	Sports associations	Social welfare, political, culture and education, national defence and peace associations
	Low	Economic and professional associations	Religious and international associations

Most often cooperation with companies concerns getting financial support for associational activities, having a common project, seeking expertise for specific tasks, organizing training or education, or providing information. Rarely have associations taken on particular tasks for companies, or vice versa. If we link these findings to questions relevant to social fields, it can be concluded that the interaction between associations and political institutions or businesses rarely resembles the 'struggles about stakes' that Bourdieu argues are a central characteristic of every 'field' of social action. Usually such social exchange occurs in relation to practical matters concerning which no 'field' can actually develop. Consequently, instead of challenging the structures of political power or established institutions, voluntary associations typically tend to adjust themselves to the rules of the game prevailing in their respective subfields.

It should not come as a surprise then that the associations' most important form of cooperation with town institutions is that which aims at securing economic resources. The fact that fund-raising functions are the principal ties between the majority of associations and local governments and enterprises suggests that associations maintain such relationships primarily to serve rational self-interests and prefer to keep this side distinct from their substantive activities. Not much is known about the influence of the donating enterprises on the activities of voluntary associations in Finland. There are also lacunae in research on the new forms of neo-corporatist exchange between public and tertiary sector organizations. The outsourcing of welfare state (or municipality) tasks to private companies and third sector organizations has been a much-discussed issue in Finland since the recession of the 1990s. This transfer of welfare responsibilities from the public to the third sector has brought in pressures for associations to adopt a managerialist model of organization, which also tends to standardize their activities (Parker 2002). One-fifth of the associations involved in the survey declared of having taken responsibility for some welfare (state) tasks that previously belonged to public sector institutions, while 7 per cent have transferred some of their own original tasks to the city. In this respect differences between types of association are significant: as many as 60 per cent of social security and welfare associations have taken tasks from the city, which gives some kind of overall idea of the direction of development for the municipal welfare system (see Luomala and Pyykkönen 2002).

This reveals a very important historical change in relations between voluntary associations and the (local) welfare state: during the formation period of the Finnish welfare state, associations produced innovative proposals and expressed new needs before the political system, which responded by developing welfare structures for example by including institutions run by associations as part of the welfare state. This is how many publicly run educational and cultural institutions were created (homes for the elderly, immigration services, etc.). Nowadays the process often also heads in the opposite direction: associations increasingly adapt their activities to meet the needs of state structures and take responsibility for certain functions that are privatized by the public sector (for example, organizing welfare services for those with substance abuse problems, care for the elderly and child welfare). Part of this process is mediated by the flow of different kinds of projects that are financed by the state and municipal institutions, EU programmes, foundations, etc.

Conclusions

The analysis of the Jyväskylä data does not reveal any general, all-embracing crisis of associational social capital. A plethora of new associations have appeared on the scene during the past ten years, and a new differentiation has taken place within the system of voluntary associations: this holds true particularly for different kinds of hobby and leisure (specific cultural and sporting) associations. On the other hand, many traditional 'multi-task' (ideological, political, general cultural and traditional) associations are struggling to survive, and many have failed and died. So far there is no convincing evidence of the capacity of new specialized associations to create bridging social capital that connects people from different classes.

With the exception of large national associations that produce social or health services and labour market organizations, the average voluntary association in Jyväskylä is relatively small and holds very little economic capital. The most important social capital (in the Bourdieuan meaning of the concept) in associations comes through members and volunteers and via organizations that are connected with their interaction networks. Both old and new associations quite often have town officials or council members on their boards, and they also cooperate with the media, companies and, of course, with other associations. This strongly suggests that the close traditional relationship between local government and civic organizations still has continuity, and the city has in fact paid attention to the tertiary sector in the planning of its future strategies. Local governmental structures are usually open to local associations and/or informal citizens' initiatives that have accepted the existing 'rules of the game'. This also demonstrates that local inter-organizational networks, which transmit social capital and enable social exchange, are still functioning at least in a formal way. The Finnish welfare system has faced great challenges since the recession of the 1990s, and many of its pillars have started to be eroded during the period of economic recovery. Many reforms concerning the Finnish welfare state (effectiveness of administration, reforms of social policy, development of pension systems) were justified with economic calculations purporting to increase the competitiveness of the Finnish industry (see Julkunen 2001). These processes may have tied some older associations (for example welfare and health associations) more closely with the public sector. Among the most recent associations, interactions with local government are not as tense as among older associations, but they have found in the media a more central partner. This might be interpreted as a reflection of the tendency to stress the importance of the individual at the expense of collectives in the new system of governance. It also tells us that the role of the media is more important to new than to older associations.

The most significant changes in the structure of the system of associations concern the changing proportions of different types of associations. The mainstream is moving away from collective, co-productive, multi-task associations towards individualized, consumptive, specialized associations. One of the most important differences between older and newer associations is that the former in general belong to central unions, whereas the latter most often stay 'wild and free'. This means that the centralized four-level organizational model created along with the Finnish nation-state is in decline. New associations are also specialized and based on ties consisting of a relatively weak commitment of their members. Individual members invest less and less of their agency into a single association, and thereby less of their personal social capital. This tendency most probably decreases the sheer volume of associational social capital as well. The dispersion and specialization of associational capital makes it increasingly difficult to estimate the consequences of the formation of these new associations in the development of social capital at the societal level. The mode of action in a significant number of new associations resembles more closely the logic of co-consumption, which is likely to make large collective 'struggles' improbable. Many studies demonstrate, or at least hint, that there is no clear connection at the individual level between acting in associations and trust as such. There is no strong evidence of the ability of the new types of associations to spread solidarity and trust between different groups and strata, or even among their own members. Hence the key question should concern the role of the new types of associations in participating in struggles – even modest ones – in various social fields.

This analysis also shows that certain central traditions of Finnish civil society connected with the institution of the registered association have continued into the new century and have been accepted, at least to some extent, even by newcomer associations such as immigrants' associations. All associations, with only very few exceptions, obey the law, which is concomitant with the act of registration. In this respect new associations are often even more obedient and consensual in their relations with the political system than those established in the past. Among these latter, there were thousands of them aspiring to change the whole social system qualitatively, even if through the activities of (registered) associations. At present, the potential of the new types of associations to break the prevailing logic or codes of the political or economic subsystem remains to be investigated.

Even though it seems that there is no causal relationship between acting in an association and trust on the individual level as such, this kind of relationship exists between the average number of association memberships and active associations, and trust on the societal level. This kind of relationship is probably the factor which most directly explains international differences in voluntary activity levels. What seems important to the development of trust is, first of all, the existence of a transparent system of voluntary associations that citizens are aware of. In such a trust-embracing situation, people do not necessarily have to be active in associations themselves; it is enough to know that there are associations that function and can therefore be trusted. This has been one of the main reasons for Finland and the other Nordic countries scoring highly in international comparative studies on trust (Rothstein 2001). In addition, a prominent and properly functional welfare (state) system makes citizens feel safe and able to trust. Regardless of some partial dismantling since the mid-nineties, Finland still has such a welfare system. Finally, it seems important to have effective, perceptible dialectics between those two systems – the state and the associations of civil society. The fact that there are no great visible confrontations between the organized civil society and the local government in Jyväskylä, could tell us about the effectiveness of Finnish local government administration.

All the data analysed in the Jyväskylä study concerning the relationship between the local government and the civil society suggests that the mutual reinforcement of their interaction still functions relatively well in this particular city. Most of all, associations still advance the interests of their members both by interacting with the city authorities and by maintaining fruitful relations with companies. Local government institutions, in turn, try to piggyback on associational resources in their strategies. How thin or thick the tie of trust connected with the functioning of associations is, and what the relative weight of the memberships in different kinds of associations is in comparison with other kinds of networks (different kinds of new tribes, Internet communities, smart mobs, etc.) at different geographical levels (municipality, region, nation, globe), remain open questions.

Another interesting question is, to what extent is it possible to build a corporate polity regime (or hegemony) on the basis of new associations that a) require only light and thin commitment from the members, b) are temporal and c) act episodically? The Finnish type of corporate regime has been built on the dual capacity of associations to reveal and organize social tensions and conflicts; and at the same time makes it possible for them to communicate with the political system, and in the last instance to help its development. The relative openness of political institutions and local government have channelled associational challenges in local government, thereby often making it possible for associations to adapt to the rules of the game and to get their issues on the agenda of municipal decision-making bodies. In practice, associations that interact closely with the local government have been very effectively integrated into the institutions of local democracy. As a general conclusion drawn from these results, it seems that the borderline between civil society and the (local) government runs inside distinct associations so that those parts which interact closely with the local government tend to function as its accessories, whilst 'rank and file actors' remain in their civil society roles. Thus we may ask along with Max Weber whether the new types of associations have any of the kind of bureaucratic structures that would make them suitable for incorporation into functions of the polity regime or local government. Moreover, the question as to whether new associations produce 'good state citizens' has not been settled yet.

However, the most important impact of the system of voluntary associations on the political, welfare and economic system comes from the power of its 'collective intellect' or total collective social capital (Hänninen et al. 2003). A well-functioning political system or local government is one in which a multitude of different voluntary associations prosper. This is a conclusion drawn by scholars applying very different theoretical approaches, from Gramsci's ideas of hegemony to Putnam's concept of social capital. Associational civil society is a source of new proposals for the development of the welfare of local democracy, an alarm watch about the deficiencies in society, sewer of the breaches in social security and buffer against threats to the continuity of the institutions of government (see Siisiäinen 1986). Innovations produced by distinct associations are almost always small, often tacit, and invisible. In a society in which tens of thousands of association operate freely, the majority seems to follow the logic of 'the wisdom of crowds' (Surowiecki 2005). Some, or even many, may make mistakes or suggest innovations of low value, but there are always also associations that are on the right track in developing new solutions or reducing societal risks. Therefore, the total impact of the masses of associational innovations, challenges and demands forms a significant element of power. As the example of the Nordic countries shows, it is important that there are both conflicting and integrative movements and associations in society. Therefore the concept (or at least a metaphor) of social fields is applicable in the analysis of this contested set of social relationships, together with the idea of associations as networks of social capital. Many impacts of the whole system of voluntary associations are unintended consequences of actions of its various parts. Active and diversified associational social capital is a precondition for both a healthy reproduction of society and for its capacity to influence cultural and political change.

Bibliography

Alanen, L. (2007), 'Lasten ja aikuisten kentät: Kenttäanalyysi sukupolvisuhteiden tutkimuksessa', in L. Alanen et al. (eds).

Alanen, L., Veli-Matti, S. and Siisiäinen, M. (eds) (2007), *Sosiaalinen pääoma ja paikalliset kentät* (Jyväskylä: Koulutuksen tutkimuskeskus).

Alapuro, R. (2005), 'Associations and contention in France and Finland: Constructing the society and describing the society', *Scandinavian Political Studies* 28:4, 377–99.

Alapuro, R. (2007), 'Associations and representation in Finland and France', paper presented at the EVA seminar, Berlin, 26–27 March.

Alapuro, R., Liikanen, I., Smeds, K. and Stenius, H. (eds) (1987), *Kansa liikkeessä* (Helsinki: Kirjayhtymä).

Antikainen, A. and Houtsonen, J. (eds) (2007), *Symbolic Power in Cultural Contexts – Uncovering Social Reality* (Rotterdam: Sense Publishers).

Baer, D. (2007), 'Voluntary association involvement in comparative perspective', in L. Trägårdh (ed.).

Bellebaum, A. (ed.) (1972), *Die moderne Gesellschaft* (Freiburgg: Basel).

Bourdieu, P. (1979), *La distinction – critique sociale du jugement* (Paris: Minuit).

Bourdieu, P. (1980), 'Le capital sociale', *Actes de la Recherche en Sciences Sociales* 31:1, 2–3.

Bourdieu, P. (1985), 'The social space and the genesis of groups', *Social Science Information* 24:2, 195–220.

Bourdieu, P. (1986), 'The forms of capital', in J.G. Richardson (ed.).

Bourdieu, P. (1992), *Les règles de l'art* (Paris: Éditions du Seuil).

Bourdieu, P. (1998), *Järjen käytännöllisyys. Toiminnan teorian lähtökohtia* (Tampere: Vastapaino).

Bourdieu, P. (1996), *The State Nobility* (Cambridge: Polity).

Bourdieu, P. and Wacquant, L. (1992), *An Invitation to Reflexive Sociology* (Cambridge: Polity).

Castiglione, D. (ed.) (2002), *Interdisciplinary Perspectives on Social Capital* (Exeter: University of Exeter Political Science Department).

Cohen, J.L. (2007), 'Civil society and globalization: Rethinking the categories', in L. Trägårdh (ed.).

Crow, G. (2002), *Social Solidarities. Theories, Identities and Social Change* (Milton Keynes: Open University Press).

Curtis, J.E., Baer, D.E. and Grabb, E.G. (2001), 'Nations of joiners: Explaining voluntary association membership in democratic societies', *American Sociological Review* 66:6, 783–805.

Deth, J. van, Maraffi, M., Newton, K. and Whiteley, P. (eds) (1999), *Social Capital and European Democracy* (London and New York: Routledge).

Deth, J. van, Montero, J.R. and Westholm, A. (eds) (2007), *Citizenship and Involvement in European Democracies. A Comparative Analysis* (London and New York: Routledge).

Dunckelmann, H. (1975), *Lokale Öffentlichkeit. Eine gemeindesoziologische Untersuchung* (Stuttgart: Schriften des deutschen Instituts für Urbanistik, Band 51).

Esping-Andersen, G. (1990), *The Three Worlds of Welfare Capitalism* (Princeton, NJ: Princeton University Press).

Giddens, A. (1994), *Beyond Left and Right: The Future of Radical Politics* (Cambridge: Polity).

Hirschman, A.O. (1977), *The Passions and the Interests. Political Arguments for Capitalism before its Triumph* (Princeton, NJ: Princeton University Press).

Hänninen, S., Kangas, A. and Siisiäinen, M. (eds) (2003), *Mitä yhdistykset välittävät* (Jyväskylä: Atena).

Kaakkuriniemi, T. and Mykkänen, J. (eds) (2007), *Politiikan representaatio* (Helsinki: Valtiotieteellinen yhdistys).

Kangas, A. (2003), 'Kolmas sektori ja kulttuuripolitiikka', in S. Hänninen et al. (eds).
Kankainen, T. (2007), *Yhdistykset, instituutiot ja luottamus* (Jyväskylä: Jyväskylä Studies in Education, Psychology and Social Sciences, Nr. 326).
Kanter, R.M. (1968), 'Commitment and social organization: Study of commitment mechanisms in utopian communities', *American Sociological Review* 33, 499–517.
Konttinen, A. (ed.) (2008), *Political Participation in Finland* (forthcoming).
Küpper, W. and Ortmann, G. (eds) (1988), *Mikropolitik, Rationalität, Macht und Spiele in Organisationen* (Opladen: Westdeutscher Verlag).
Laclau, E. and Mouffe, C. (1985), *Hegemony and Socialist Strategy: Towards a Radical Democratic Politics* (London: Verso).
Liikanen, I. (1995), *Fennomania ja kansa* (Jyväskylä: SKS).
Luhmann, N. (1984), *Soziale Systeme* (Frankfurt am Mein: Suhrkamp).
Luhmann, N. (1988), 'Organisation', in W. Küpper and G. Ortmann (eds).
Luomala, A. and Pyykkönen, M. (2002), 'Yhdistysten yhteistyökumppanit ja kanssakäymisen muodot', in M. Siisiäinen (ed.).
Maloney, W.A. and Rossteutscher, S. (eds) (2007), *Social Capital and Associations in European Democracies. A Comparative Analysis* (New York: Routledge).
Melin, H. and Nikula, J. (eds) (1993), *Yhteiskunnallinen muutos* (Tampere: Vastapaino).
Morales, L. (2001), 'Citizens in polities: The individual and contextual determinants of political membership in Western Countries', *Working Paper 2001/164* (Madrid: Intituto Juan March de Estudios e Investigaciones).
Morales, L. and Geurts, P. (2007), 'Associational involvement', in J. van Deth et al. (eds).
Möttönen, S. and Niemelä, J. (2005), *Kunta ja kolmas sektori. Yhteistyön uudet muodot* (Jyväskylä: PS-Kustannus).
Negt, O. and Kluge, A. (1976), *Öffentlichkeit und Erfahrung. Zur Organisationsanalyse von bürgerlicher und proletarischer Öffentlichkeit* (Frankfurt am Mein: Suhrkamp).
Neuendorf, H. (1973), *Der Begriff des Interesses. Eine Studie zu den Gesellschaftstheorien von Hobbes, Smith und Marx* (Frankfurt am Mein: Suhrkamp).
Nummela, I. (ed.) (1997), *Jyväskylän kirja* (Jyväskylä: Jyväskylän kaupunki).
Nyyssönen, H. (ed.) (1993), *Nimet poliittisessa retoriikassa*, Valtio-opin laitos, Julkaisuja 66 (Jyväskylä: Jyväskylän yliopisto).
Ojanen, E. (1992), 'Urheiluseurojen nimet 1–10', *Helsingin Sanomat*, feature article series 2–23 July.
Oliver, P. (1984), '"If you don't do it, nobody else will": Active and Token Contributors to Local Collective Action', *American Sociological Review* 49, 583–88.

Opp de Hipt, M. (1987), *Denkbilder in der Politik* (Opladen: Westdeutscher Verlag).
Palonen, K. (1990), 'Benämning som politisk akt. Reflexioner kring namnpolitiken', paper presented in the Working group 'Det politiska språket, NOPSA-Kongressen, Reykjavik, 16–18 August.
Parker, M. (2002), *Against Management: Organization in the Age of Managerialism* (Cambridge: Polity).
Pekonen, K. (1993), 'Nimet poliittisessa symboliikassa', in H. Nyyssönen (ed.).
Piirola, H. (1985), *Suomen Valtakunnan Urheiluliiton urheiluseurojen nimistö 1865–1980*, Suomen kielen pro gradu-tutkielma (Jyväskylä: Jyväskylän yliopisto).
Putnam, R.D. (1993), *Making Democracy Work. Civic Traditions in Modern Italy* (Princeton, NJ: Princeton University Press).
Putnam, R.D. (1995), 'Bowling alone: America's declining social capital', *Journal of Democracy* 6:1, 64–78.
Pyykkönen, M. (2002), 'Vapaaehtoistyö jyväskyläläisissä yhdistyksissä', in M. Siisiäinen (ed.).
Richardson, J.G. (ed.) (1986), *Handbook of Theory and Research for the Sociology of Education* (New York: Greenwood Press).
Richter, R. (1985), *Soziokulturelle dimensionen freiwilliger Vereinigungen* (München: Beiträge zur Kommunalwissenchaft).
Rothstein, B. (2001), 'Social capital in the social democratic welfare state', *Politics and Society* 29:2, 207–41.
Rothstein, B. and Dietlind, S. (2002), 'Social capital, impartiality and the welfare state: An institutional approach', in D. Castiglione (ed.).
Ruuskanen, P. (2001), *Sosiaalinen pääoma – käsitteet, suuntaukset ja mekanismit*, VATT tutkimuksia 81 (Helsinki: VATT).
Schofer, E. and Fourcade-Gourinchas, M. (2001), 'The structural contexts of civic engagement: Voluntary association membership in comparative perspective', *American Sociological Review* 66:6, 806–28.
Siisiäinen, M. (1986), *Intressit, yhdistyslaitos ja poliittisen järjestelmän vakaisuus*, Jyväskylä Studies in Education, Psychology and Social Research 57 (Jyväskylä: Jyväskylän yliopisto).
Siisiäinen, M. (1988), 'Järjestöllinen pääoma: käsite ja merkitys yhdistystutkimuksessa', *Hallinnon tutkimus* 7:3, 154–69.
Siisiäinen, M. (1990), *Suomalainen protesti ja yhdistykset* (Jyväskylä: Tutkijaliitto).
Siisiäinen, M. (1992), 'Social movements, voluntary associations and cycles of protest in Finland 1905–1991', *Scandinavian Political Studies* 15:1, 21–40.
Siisiäinen, M. (1999), 'Voluntary associations and social capital in Finland', in J. van Deth et al. (eds).
Siisiäinen, M. (2002a), 'Yhdistyslaitoksen yleiskuva Jyväskylässä', in M. Siisiäinen (ed.)

Siisiäinen, M. (2002b), *Yhdistykset kolmannen sektorin toimijoina. Tutkimus jyväskyläläisistä yhdistyksistä* (Helsinki: Sosiaali- ja terveysturvan keskusliitto).

Siisiäinen, M. (2003a), 'One concept two approaches: Bourdieu and Putnam on social capital', *International Journal of Contemporary Sociology*, 40:2, 183–203.

Siisiäinen, M. (2003b), 'Vuoden 1997 yhdistykset', in S. Hänninen et al. (eds).

Siisiäinen, M. (2003c), 'Muuttuvat yhdistykset', in H. Melin and J. Nikula (eds).

Siisiäinen, M. (2007a), 'Uudet yhdistykset ja intressien representaatio', in T. Kaakkuriniemi and J. Mykkänen (eds).

Siisiäinen, M. (2007b), 'Symbolic power as a critical concept', in A. Antikainen and J. Houtsonen (eds).

Siisiäinen, M. (2008), 'Differentia specifica of voluntary organizing in Finland', in A. Konttinen (ed.).

Simmel, G. (1968), *Soziologie. Untersuchungen über die Formen der Vergesellschaftung*, 5. Auflage (Berlin: Duncker & Humbolt).

Stenius, H. (1987), *Frivilligt, jämlikt, samfällt*, Skrifter utgivna av Svenska Literatursällskapet i Finland, 454 (Ekenäs: SLF).

Strømsnes, K. and Wollebæk, D. (2006), 'Mapping voluntarism: Scandinavia in comparison', paper presented at the International ISTR Conference, Bangkok, 9–12 July.

Surowiecki, J. (2005), *The Wisdom of Crowds* (New York: Anchor Books).

Tenbruck, F.H. (1972), 'Gesellschaft und Gesellschaften', in A. Bellebaum (ed.).

Tenbruck, F.H. (1989), *Die kulturellen Grundlagen der Gesellschaft: der Fall der Moderne* (Opladen: Westdeutscher Verlag).

Touraine, A. (1988), *Return of the Actor: Social Theory in Postindustrial Society* (Minneapolis: University of Minnesota Press).

Touraine, A. (2000), *Can We Live Together: Equality and Difference* (Cambridge: Polity).

Trägårdh, L. (2007), 'The "civil society" debate in Sweden: The welfare state challenged', in L. Trägårdh (ed.).

Ursin, H. (1990), *Järjestöjen aluerakenne Suomessa.* Kulttuurimaantieteen pro gradu-työ (Helsinki: Helsingin yliopisto).

Wallace, C. (2005), *Trends in Social Capital in the EU* (Aberdeen: University of Aberdeen).

Warren, M.E. (1999), 'Democratic theory and trust', in M.E. Warren (ed.).

Warren, M.E. (ed.) (1999), *Democracy and Trust* (Cambridge: Cambridge University Press).

Weber, M. (1911), 'Geschäftsbericht', in *Verhandlungen des ersten deutschen Soziologentages vom 19–22. Oktober* (Tübingen: J.B.C. Mohr).

Weber, M. (1976), *Wirtschaft und Gesellschaft* (Tübingen: J.B.C. Mohr).

Wollebæk, D. and Selle, P. (2002), *Det nye organisasjonssamfunnet: demokrati i omforming* (Bergen: Fagbokforlaget).
World Bank (2005), *World Development Indicators 2005* (Washington, DC: World Bank).

Index

For Product Safety Concerns and Information please contact our EU representative GPSR@taylorandfrancis.com
Taylor & Francis Verlag GmbH, Kaufingerstraße 24, 80331 München, Germany

www.ingramcontent.com/pod-product-compliance
Ingram Content Group UK Ltd.
Pitfield, Milton Keynes, MK11 3LW, UK
UKHW020017200926
475702UK00014B/1933
9781138245457